INSTRUCTOR'S MANUAL

to accompany

BRIEF CALCULUS

for

BUSINESS, SOCIAL SCIENCES, AND LIFE SCIENCES

preliminary edition

DEBORAH HUGHES-HALLETT
Harvard University

ANDREW M. GLEASON
Harvard University

PATTI FRAZER LOCK
St. Lawrence University

DANIEL FLATH
University of South Alabama

et al.

Prepared by

ANSIE MEIRING
University of Pretoria

LYNN E. GARNER
Brigham Young University

DANIEL FLATH
University of South Alabama

JOHN WILEY & SONS, INC.
NEW YORK • CHICHESTER • WEINHEIM • BRISBANE • SINGAPORE • TORONTO

This project was supported, in part,
by the

National Science Foundation

Opinions expressed are those of the authors
and not necessarily those of the Foundation

Grant No. DUE-9352905

ISBN 0-471-17660-5

Printed in the United States of America

10 9 8 7 6 5 4 3 2 1

Printed and bound by Victor Graphics, Inc.

PREFACE

The best way to get a feel for what has been done in the preparation of *Brief Calculus for Business, Social Sciences, and Life Sciences* is to read the book and tackle the problems. If you wish to read the Instructor's Manual first, here it is. The first part gives an overview of the book, the general approach and philosophy, and the scope. The second part comments on each section of each chapter and suggests a representative set of problems. We recommend that you read Part One entirely and use Part Two as a resource as you prepare individual classes.

Part Three offers sample syllabi for help in designing the course; Part Four provides overhead masters for aid in presentation of particular topics. Part Five gives calculator programs or procedures for accomplishing various tasks for which the technology is particularly useful; since not all calculators have all features built in, programs are sometimes necessary. A case in point is numerical integration; while each calculator can give its own numerical estimate of a definite integral, if you want left and right hand sums or the trapezoidal estimate you must use your own programs. Part Six provides sample exam questions.

Table of Contents

PART I

THE BOOK

The aim of *Brief Calculus for Business, Social Sciences, and Life Sciences* is to present basic calculus in a way that

- demands greater understanding and less routine manipulation
- covers less material in greater depth
- presents concepts graphically, numerically, and verbally as well as algebraically
- develops concepts from common sense investigations rather than abstract definitions
- incorporates the appropriate use of technology
- is written for students, to be read by students
- uses meaningful applications

Greater Understanding, Less Manipulation

Students are expected to understand the concepts of derivative and integral beyond the purely symbolic level; in return they are not asked to perform great feats of symbolic manipulation.

For example, rather than requiring the memorization of δ-ϵ definitions or proofs, the student is asked to demonstrate real understanding of the difference quotient that underlies the definition of the derivative. Rather than emphasizing techniques of integration, the student is required to understand why the definite integral of a rate of change is the total change.

Throughout the book, students are asked to think through a situation and investigate it for themselves. Applications are not organized into template problems. They require of students less dependence on recipes and more prolonged thought than traditionally expected.

Less Material, Greater Depth

The following have been omitted:
- introductory chapters on precalculus material
- the formal definition of limit; limits as a separate topic
- algebraic techniques of integration

Further,
- curve sketching has been replaced with an emphasis on graphs and the qualitative behavior of parameterized families of curves
- there is less emphasis on theorems
- definite integrals are usually computed numerically

In return, we expect much more thorough understanding of the key concepts of derivative and definite integral, spending more time in developing these concepts and requiring more thought and communication on the part of the student.

The Rule of Four

The Rule of Four is that every concept should be introduced graphically, numerically, analytically, and verbally. This means a lot more than drawing more graphs or computing a few difference quotients; it means giving equal time to all four ways of understanding concepts. It means expecting and accepting graphical explanations; it means acknowledging the validity of numerical calculations properly done; it means requiring students to communicate their understanding in practical terms, such as colleagues or employers without calculus training might understand. We still expect students to carry out the correct algebraic manipulations that might be necessary, but we accept graphical and numerical reasoning as well.

The Rule of Four demands a few changes in approach:
- numerical methods such as approximating derivatives, estimating integrals, and fitting curves to data become essential topics

- calculators and computers make possible a shift in emphasis from constructing to interpreting graphs
- student assessment must include verbal explanations

The Way of Archimedes

The Way of Archimedes is the philosophical stance that definitions and proofs evolve from a long process of common sense investigation, rather than starting with abstract definitions. It is the attitude that students learn best when they develop the concepts the same way that researchers develop concepts, rather than being given an abstract structure and deducing its properties in a purely theoretical way. In the words of Archimedes:

> ...I thought fit to write out for you and explain in detail ... the peculiarity of a certain method, by which it will be possible for you to get a start to enable you to investigate some of the problems in mathematics by means of mechanics. This procedure is, I am persuaded, no less useful even for the proof of the theorems themselves; for certain things first became clear to me by a mechanical method, although they had to be demonstrated by geometry afterwards because their investigation by the said method did not furnish an actual demonstration. But it is of course easier, when we have previously acquired, by the method, some knowledge of the questions, to supply the proof than it is to find it without any previous knowledge. [1]

In the spirit of Archimedes, the mathematician looking over the shoulder of the author is not the focus of our attention. We do not attempt to state and prove most of the theorems of elementary calculus; the justifications that are given, though mathematically correct, are conceptual rather than formally rigorous, for we expect students to read and understand them. Functions are introduced in the context of real-life uses rather than abstractly and rigorously.

Real numbers are often represented by decimal approximations in this text; decimals are more appropriate in most applications than exact values: the answer 4.443 is often more useful than than $\pi\sqrt{2}$. Most importantly, it is desired to give students the freedom to use their common sense.

Technology

This text assumes that students have access to computer programs or graphing calculators that can be used to

- estimate a zero of a function
- draw the graphs of a function in an arbitrary viewing window
- estimate a definite integral

It is helpful (but not essential) to have access to a computer or calculator that will

- estimate the derivative of a function at a specified point
- draw a contour diagram for a function of two variables
- compute a regression line or exponential regression function for a set of data

Programs for several models of calculators are given in Chapter Five of this manual.

The advantage of using technology is that it can enable students to experiment, make conjectures, and verify special cases of general results. Inappropriate uses of the technology are those that make it a crutch. It should not be used when the desired result is more quickly obtained mentally. There are no icons telling students to use a calculator or computer, as we believe that students should learn to decide when to use technology and when not to use it.

A Book to be Read

This text is neither a reference book nor an encyclopedia, but a presentation of the basic ideas of calculus in an applied context. It was written with students in mind, informally so that students will

---

[1] From The Method, in The Works of Archimedes, ed. and trans. by Sir Thomas L. Heath (New York: Dover)

feel comfortable reading it. It has not been written to facilitate what students usually do first, look at homework problems, then look for worked examples that fit the template, and finally as a last resort begin to read the text. It is well to warn students that this text will be different.

The Problems

The problems are the heart of the text. There are fewer rote problems than in many texts, and more problems requiring careful thought and the ability to write clearly. There are some essay questions, questions that require graphical and numerical work rather than algebra, and questions that require the use of technology. Assign fewer problems than you have been accustomed to, and give the students more time.

Verbal explanations are an essential part of the course. Composing written answers helps students clarify their own understanding and reading these explanations will provide you with valuable insight into the students' thinking processes.

Many of the problems in the text do not have unique correct answers. If you use student readers or lab assistants, urge them to be open to a variety of interpretations on these problems. Be flexible. For such problems, any reasonable answer, with reasonable justification, should be acceptable.

Although most of the problems are not difficult, students are not accustomed to reading and interpreting math problems nor to modeling real-world situations. They expect to see problems that mimic the examples in the text and have unique answers in the back of the book. The initial shock at finding these conditions changed will have to be dealt with. You may need to reassure students frequently that this new approach is difficult at first but that their honest efforts will pay off handsomely in the end.

Overview by Chapters

Chapter 1 introduces the key properties of the most basic functions used in the text: linear; exponential, and power functions. Though the functions themselves should be familiar, their graphical and numerical properties and practical uses usually are not. Functions are couched in the context of understanding real-world processes. Functions represented by tables of data receive particular attention, including an introduction to curve fitting.

Chapter 2 discusses the key concept of the derivative according to the Rule of Four. Differentiation formulas are deliberately avoided in order to better develop the mathematical ideas.

Chapter 3 discusses the key concept of the definite integral in the same spirit as Chapter 2. It is possible to delay Chapter 3 until after Chapter 5.

Chapter 4 introduces additional functions useful in mathematical modeling, including the logarithm, sine and cosine. The emphasis throughout is on applications.

Chapter 5 presents the symbolic approach to differentiation.

Chapter 6 presents applications of the derivative and definite integral. The Way of Archimedes is used to investigate curves and applications to the social sciences. Techniques of integration are omitted in favor of applications; technology is used to estimate the definite integrals that result.

Chapter 7 introduces functions of two or more variables using contour diagrams, tables and formulas, followed by the elementary differential calculus of functions of several variables.

The First Three Chapters

The first three chapters of the text illustrate the spirit of the work. It is important not to go too rapidly through these chapters.

We know from experience how important it is for students to have a thorough grasp of the material. Although much of the material may be familiar to the students, it is presented in a sufficiently different manner that most will find it quite new. It is crucial that students become comfortable with graphical and numerical work early on. The idea of Chapter 1 is to make students familiar with functions from every point of view. For example, a student should be able to identify the family to which a function belongs from its graph alone, to identify a linear function by looking at a table of values, and to describe in graphical terms the relative growth rates of functions.

Chapters 2 and 3 introduce the key concepts of derivative and definite integral one after the other so as to show clearly the relation between them. Only after this introduction do we get into techniques and formulas for derivatives and integrals. You should cover these chapters slowly enough that students have time to think about what the key concepts really mean.

Where Students Have Difficulties

Students are often reasonably proficient at what is usually emphasized in mathematics courses: using rules to manipulate formulas. They are much less proficient at understanding and interpreting mathematics critically, as well as applying it to practical situations. This means that many students initially find the material in this book difficult. In particular, if you have students who have done well in the past in courses emphasizing manipulation, both you and they may be surprised at the difficulty they have. Dealing with their apprehension will probably require your repeated reassurance, particularly early on.

Many students have much more difficulty with tables and graphs than they do with formulas. They confuse functions with their formulas, the rules that generate them. Tables are usually foreign to them; they need time and practice to get used to them. Graphs can also be difficult; while most students will have seen graphs before, many students will have difficulty interpreting them. For example, some students will have difficulty interpreting a graph of distance versus time, confusing it with a trajectory. A graph of velocity vs. time can also cause them trouble. This difficulty with interpretation is one of the reasons for the emphasis on interpreting the derivative and the definite integral. Bear in mind that the students in your class who have been weak in algebraic skills may welcome the opportunity to learn calculus with the assistance of graphs and tables.

Students often have considerable difficulty thinking geometrically. Many, for example, cannot read the slope of a line from its graph. Try to wean them away from dependence on formulas to estimate such quantities. They may also have difficulty interpreting geometric objects; many confuse the secant line itself with the average rate of change (which is the slope of the secant line) or the tangent line with the derivative (which is the slope of the tangent line).

Many students will have difficulty with basic material, such as exponential functions and logarithms, or even percents and fractions. We suggest that you not spend too much time at the beginning of the course going over manipulative rules but instead review as you go along whenever it is needed.

Students may also misinterpret the results on their graphing calculators or computers. This is often due to the computer's tendency to deal badly with the very large (say, near an asymptote) and the very small (roundoff error). Finding an appropriate viewing region for a graph can be difficult for students as well. They often give up too easily when graphing to find points of intersection, especially when large values are involved (for example, $y = e^x$ and $y = 2 + 100x$). Students sometimes also assume that if the cursor is on top of the point of intersection then the (x, y) values are exact, whereas these values can actually be off by large amounts, depending on the window settings.

Teaching Style

In the spirit of the book, try to make your class interactive; make it a class rather than a lecture. Since many of the problems can be done by several different methods, a useful class discussion can result from having students explain to the rest of the class the method they used and why. As you develop new ideas, ask questions to the class (and give them time to answer) to get students into the habit of thinking ahead and asking themselves "what if." Encourage interaction among the students, even outside of class; suggest that their classmates can help them figure out if they have done the problems satisfactorily.

The philosophy of this book is that mathematics is not a set of formulas, rules, and recipes, but rather a process of making sense out of the world we live in. Mathematics is a way of understanding what we observe, and in this sense it richly deserves the name "natural philosophy" that it had at the time of Newton.

PART II

THE CHAPTERS

This part of the manual contains more detailed information about each chapter and section of the book. If you have trouble knowing what to do with a particular section or deciding which points to emphasize, look at the information on that section. There you will find information on timing (50-minute periods are assumed), suggestions on how to introduce the topic, extra examples, tips on what students find hard, ideas for computer or calculator use, and a discussion on the problems in each section. Note that you may not be able to cover every section of the book carefully in one semester, so be prepared to make some choices.

Suggested problems are selected to reflect the Rule of Four, wherever possible. We do *not* recommend that you assign all the suggested problems; you may need to adjust the number upward or downward depending on the needs of your class and the speed at which you intend to cover the material.

CHAPTER ONE

Eight or nine classes.

Overview

This chapter sets the background and the tone for the whole book. It is a book about mathematical modeling of real world phenomena. From the first page there is a focus on things that change and their rate of change. The topics in the first chapter are all precalculus topics, but the graphical, numerical, and modeling approach is probably new to most of the students. Keep the flow lively and try not to get bogged down in any particular topic, but don't skip anything, either.

The most basic functions used in this book are introduced in this chapter, the linear, exponential, and power functions. (Chapter 4 introduces a second round of functions, including the natural logarithm, sine, cosine, and others useful in modeling.) Our purpose is to acquaint the student with each function's individuality (its graphical shape, identifying properties, comparative growth rates, and general uses), to train students to read graphs and think graphically, to read tables and think numerically, and to apply these skills to the mathematical modeling of the real world. Students are being asked to think about mathematics in new ways. Be assured, and assure them, that time and effort spent in this chapter will pay off well in subsequent chapters.

1.1 HOW DO WE MEASURE CHANGE?

One class.

Key Points

Total change and average rate of change. Δ-notation.

Ideas for the Class

Students will want to know from the first day how calculus will be relevant to their own goals, which for students of this course are not fundamentally mathematical goals. Suggest to them that calculus is the mathematical basis for the careful study of things that change over time. They will be able to name in class many examples of changing phenomena that interest them.

Students understand immediately that the first mathematical step in an analysis of many of their examples is to quantify the change. They may be surprised to find that there are two ways to do it. In this section you want to bring the students to an appreciation of the distinction between total change and (average) rate of change, using real world examples that they already have some familiarity with. Nothing could be simpler than the lead-off example in the text on the growth of a child, and yet it shows the distinction very clearly. Point out that change and rate of change are measured in different units, in the growth example in 'inches' and in 'inches per year' respectively. Students may notice the analogy with the familiar concepts of distance and velocity.

The Δ-notation is one of the simplest in common use by mathematicians. Tell the students that Δ stands for "difference" and make it a point to use the notation in many of your examples. Total change in a debt D is a *difference* ΔD and rate of change of the debt is a *difference quotient* $\Delta D / \Delta t$, as in Example 1.

It is also important to be able to interpret a statement as an average rate of change or as a total change. For example: "The concentration of salt in the water changed during the first five minutes by 10 mg/ml." IS this rate of change or total change? Once they have identified this as total change, ask them to rephrase this as rate of change.

Problems

Problems 1, 3, 5, 6 & 7 all involve data presented in a table. Students might take a little time to become familiar with this type of representation. They would benefit greatly by doing a number of these.

Problem 2 requires of the students to find approximate values from a graph. The problem itself guides the student in finding the required rate of change. This is the only problem of this kind in this set of problems.

Problem 4 is a good problem for generating a discussion. The students could be asked to add to this interesting list of natural phenomena for the same question.

In Problem 8 the information is not explicitly given in a table. This complicates the problem somewhat. This type of formulation encourages careful reading. It is a good problem to assign.

Suggested Problems

#1, 2, 4, 7, 8

1.2 WHAT'S A FUNCTION?

One class.

Key Points

The Rule of Four: algebraic, graphical, numerical, and verbal descriptions of functions. Supply and demand curves and equilibrium price. Proportionality.

Ideas for the Class

Many students at this stage, depending on their backgrounds, regard functions as formulas. Emphasize the much more general concept of a function as a rule that uniquely assigns one number to another number, and lead off with numerical and graphical examples of defining a function, in accordance with the Rule of Four.

For example, you might start with the following table:

TABLE 1.2.1

t	-3	-2	-1	0	1	2	3
$f(t)$	16	96	144	160	144	96	16

Then represent the data graphically, and ask students to come up with a formula for the function. Then indicate that $f(t)$ represents the height above ground, in feet, of an object thrown into the air so that its highest point is reached at time $t = 0$, in seconds. Have students discuss salient features of the table, the graph, and the formula $f(t) = 160 - 16t^2$, noticing how those features are reflected in the other representations of the function. For example, the maximum value in the table is 160; see how that fact is supported in the graphical, algebraic, and verbal descriptions of the function. Tell them that we want them to learn how to read the same properties of a function from any of its representations. Functions that arise naturally from real situations have more appeal as the Way of Archimedes claims, if you can find them.

It may take some students a bit of time to get comfortable with the use of function notation, so it does not hurt to begin with extremely simple examples. Make sure that your students fully understand that the expressions for change and average rate of change of a function given in this section are just different ways of describing the same concepts from the preceding section. One way to do this might be to quickly revisit an earlier in-class example, this time using the language of functions.

Supply and demand curves provide a natural opportunity to discuss functions graphically. They will make many further appearances in the text, so it is a good idea to treat them explicitly in class now.

Problems

Problems 1-7 all deal with the graphical representation of a function. Be sure to be flexible on the answers. The first few of these are quite simple but Problems 6 & 7 require careful thinking. Assign at least one of these.

Problem 8 is a good problem to assign. Students should feel comfortable with it because of the algebraic formulations. Students should also feel very comfortable with Problems 11 & 12 because of the algebraic representation. It won't be necessary to assign more than one of these.

Problems 13-16 as well as Problems 19 & 20 deal with supply and demand - a new topic for most students. They need to do a number of these problems.

Problems 17, 18, 21 and 22 all deal with tables of data. Students should become more and more familiar with tables of data.

Suggested Problems

#1, 2, 5, 6, 8, 12, 13, 14

1.3 LINEAR FUNCTIONS

One class.

Key Points

Linear functions have the same average rate of change over every interval. Equivalently, a table of values with regularly spaced entries will show constant differences. Slope as average rate of change. New vocabulary: increasing, decreasing, family of functions, parameter.

Ideas for the Class

Follow the Rule of Four. Start with a table of values and ask the students to give the slope or the formula. It is best to use real data such as the pole vault or search-and-rescue examples in this section. Otherwise, students think you are just torturing them by creating a table and then hiding the formula that created it. Have the class focus on how the formula comes from the table, not the other way around. Students may have trouble at first realizing that they can read the slope as a change in the y-value divided by the corresponding change in the x-value. You might want to include an example in which the x-values are not evenly spaced.

Also, give students a verbal description of a linear function and ask them to come up with a formula. For example: The cost of a case of apples fresh from the orchard is $20, but drops by 35 cents per day; write a formula for the cost of a case of apples that has been sitting in the store for t days. Again emphasize that the characteristic property of a linear function was embodied in this verbal description in the constant rate of decrease in price.

Students must see that slope is just another way of describing the theme topic of the course, which is rate of change. Given a linear function described either numerically or graphically, ask first for an average rate of change for the function over an interval, and then ask for the slope. If the students laugh and say they are the same question, you will know you have succeeded.

Keep in mind that in real applications the two axes of a graph are usually scaled quite differently; even the units of measurement are often different. Try to reflect this in your choice of examples, say on occasion marking off the x-axis in units of 20 and the y-axis in units of 10,000. Students should be able to write the formula for any linear function given its graph. If you want to calculate some slopes from points, have different students use different pairs of points and point out that the slope is independent of the pair of points chosen.

The terms increasing and decreasing are introduced in this section. Emphasize that these terms apply as the independent variable increases, moving from left to right. Some students want to say that the function $y = x^2$ is increasing for negative x, and the graph does rise as you move away from the origin in a negative direction, but point out that this is not what the definition means.

This section includes the first use of the terms *family of functions* and *parameter*, both of which are fundamental concepts in modeling situations. Encourage students from the first to think in terms of families of functions rather than of individual functions.

Problems

The first few problems involve formulas for linear functions. The rest mainly involve graphical and numerical interpretations - concentrate on these.

Problems 1-5 are reminiscent of earlier courses; students will probably feel comfortable with these. For Problem 6, be flexible with the answers you get, for the slope is estimated.

Problems 7-10, 13, and 15-17 all involve getting a linear function from a table of data. The students would benefit by doing more than one of these.

Problems 19-22 involve the graphical interpretation of average rate of change. Problem 21 is good for developing the skill of graphical interpretation and is a good discussion problem.

Problems 24-29 might seem slightly difficult to the students. The linear function involved might seem obscured. Important skills are developed here - that of reading and interpreting.

Suggested Problems

#1, 6, 7, 8, 17, 19, 24, 25

1.4 APPLICATIONS OF FUNCTIONS TO ECONOMICS

One class.

Key Points

Introduction of the terms cost, revenue, profit. Depreciation, budget constraints, and supply and demand curves again, mostly linear, with effects of taxation. Vocabulary: fixed cost, variable cost, break-even point.

Ideas for the Class

Many of the ideas in this section may be familiar to the students, and even if not they are intuitively grasped. This section illustrates the use of linear functions in economics.

Work out some simple, common-sense examples of cost/revenue/profit, straight-line depreciation, budget constraints, and equilibrium points in supply/demand. Students may have difficulty thinking through the effects of taxes on supply/demand equilibrium, so a couple of examples are in order. In particular, if the tax is paid by the producer, the effect is to reduce the supply and drive up the equilibrium price, but if the tax is paid by the consumer, the effect is to reduce the demand and drive down the equilibrium price. You might get students to discuss these facts from a purely graphical point of view, noting how the tax changes the supply or demand curve. In each case, the equilibrium supply is lowered, and both producer and consumer end up sharing the tax burden.

It is mentioned that even though many applications require cost and revenue functions to have discrete domains (whole-number units q), we are treating them as though they were continuous and smooth and defined for all $q > 0$. If you have a class of economics students, they may be familiar with the convention of considering the cost function $C(x)$ as the cost in dollars per week of producing x units per week rather than as the cost in dollars of producing x units, a convention in which the independent variable is naturally continuous rather than discrete.

Problems

These problems aim to cultivate two skills: that of drawing up cost, revenue, supply and demand functions and that of analyzing such given functions. Again, students are more familiar with algebraic representations.

Problems 3, 8, 9 and 11 deal with numerical and graphical representations and are good problems to assign.

Problem 10 is a discussion problem.

Problems 12-18 all require drawing up equations. The more interesting ones are problems 14 (depreciation) and 16 & 18 (budget constraints).

Problems 20 and 21 deal with sales tax. This is an interesting but difficult concept. A good idea would be to draw graphs for explaining what happens to equilibrium prices. A discussion should then lead to problem 22 which deals with symbols rather than numbers.

Suggested Problems

#1, 2, 3, 8, 11, 14, 16, 20, 21

1.5   EXPONENTIAL FUNCTIONS

Two classes.

Key Points

After linear functions, exponential functions are the most important for applications. The values of these functions change by constant ratios over equal intervals. Contrast with the way linear functions change. Growth and decay rate. Doubling time and half life. Concavity. Relative rate of change.

Ideas for the Class

Take your time with this section. Most students are not nearly as familiar with exponential functions as with linear ones and this fundamental section may not seem like review to them.

Start with a table of values for a simple exponential function, and ask the students how they know it is not the table of a linear function. Then let them notice that the ratios of consecutive values are constant, and ask them to find the formula for the exponential function. Bring in doubling time, too.

Then take a look at the population of Mexico as in the text. Students sometimes wonder how experts make population predictions, and so are usually pleased to be let in on one of the simplest of methods, which is based on the fact that many populations grow exponentially. Perhaps you could ask them to find some data for other countries or even states or cities and report to the class on whether growth was exponential, and if so what the growth factor and what they predict for twenty years into the future. Spend a little time with population growth now, especially if you plan to discuss section 4.5 on fancier growth models later in your class. It is a good idea at this point to mention the dangers of extrapolating too far into the future.

Then discuss decreasing exponential functions, and characterize gowth and decay by the size of the base. Talk about half-life, too. Note the distinction between growth factors and growth rates; if $P = P_0(1 + r)^t$, then r is the growth rate over one time period, and $1 + r$ is the growth factor, and if $P = P_0(1 - r)^t$, then r is the decay rate over one time period, and $1 - r$ is the growth (or decay) factor. Later (in Chapter 4), we will want to contrast r with *continuous* growth and decay rates, which are instantaneous rates.

Take some time to discuss exponential functions as the second *family* of functions in the students' toolbox. (The first is the family of linear functions.)

Concavity is introduced in this section. Be sure to give examples of increasing and decreasing functions with all types of concavity (including none). An in-class exercise that both amuses and instructs the students is to draw a circle on the board and ask them to find sections of it that exhibit each of the four possible combinations of increasing/decreasing and concave up/concave down.

It is very natural for students to want to discuss percentage increase in a population. Ask them if the percent increase is a change in the population or a rate of change in the population. In fact it is neither - since it is not computed by either a difference or a difference quotient. It is a relative rate of change, a new concept introduced in this section.

Problems

Problems 1-4 cultivate a desired way of thinking. Students will benefit from doing at least two of these.

Assign at least one of problems 5-8 and discuss the "why" aspect thoroughly.

Problems 9-12 involve tables of data with which students should enjoy working by now. Problem 12 involves the issue of "approximately exponential". Plotting the points might help the students to become more familiar with this notion.

Problem 13 is simple but reinforces the link between exponential growth and constant percentage growth.

Exponential decay is normally a more difficult concept than exponential growth. Problems 14 and 15 deal with this.

Problem 18 is an excellent problem because it deals with both linear and exponential growth. Be sure to assign it. Problem 20 is a good discussion problem. Students like analyzing real-life quotes.

In Problem 21 the students get a chance to draw up a table themselves and also to make sure that they know the difference between absolute and relative growth rates. Problem 22 encourages graphical understanding and is a good problem to assign.

Problem 23 touches upon the idea of making predictions.

Problems 24 and 25 are for students who are on top of the work and would like to do more.

Suggested Problems

#1, 3, 5, 11, 12, 13, 18, 21, 22

1.6   POWER FUNCTIONS

One class.

Key Points

Power functions and their graphs. Comparison with exponential functions.

Ideas for the Class

This is an opportunity to let the students work on graphing calculators or computers while you wander around. They should draw lots of graphs in various windows to discover the behaviour of power functions. For example, sketch graphs of x^2, x^3, x^4, x^5, etc., first in a window such as $-5 \le x \le 5$ and $-10 \le y \le 10$, and then in $0 \le x \le 1$ and $0 \le y \le 1$, and then in $0 \le x \le 5$ and $0 \le y \le 100$. Then have them sketch graphs of power functions with fractional exponents in similar windows, then negative powers. Many students have trouble 'remembering' the shapes of graphs of functions such as $x^{2.5}$. Here is a chance for them to benefit from thinking in terms of the entire family of power functions rather than just the one function that baffles them, for if they think 'family' they will realize that the graph they seek fits between the graphs of x^2 and x^3 which they know very well.

If you do not have the facilities for students to do this work themselves, try to have a computer or calculator on which you can demonstrate the above activities. Lacking that, prepare transparencies illustrating how the graphs change when the windows change.

A particularly nice exercise for classroom use or group work is Problem 22. The principle, of course, is that any increasing exponential will eventually outrun every power function. Different windows on the graphs, however, might suggest a different conclusion; have students reason about the various windows and make suggestions as to the window parameters for each picture that you generate.

Don't be afraid to take time with these activities, nor to be too elementary. Students typically have had little experience with varying window parameters, and may surprise you or themselves with the effort it takes to catch on. Don't underestimate how hard it can be for students to choose a good window on a computer or calculator.

Problems

Most of these problems are intended to be tackled with the aid of technology. Although some of them can be done by hand, the results appear more slowly and students will have difficulty in making connections.

Problems 1-5 are simple exercises for students to start off with.

Problem 6 should be assigned. It is excellent for cultivating insight.

Problems 7 & 8 are discussion problems - "why" is important. Problems 21 & 22 are also in the "why" category, somewhat more difficult but equally valuable.

Problem 9 should be assigned because it follows the thread of the general idea in the textbook.

Problems 11-18 are simple but constructive problems. Assign at least two of these.

Problem 19 could be a somewhat difficult problem to students, but should be assigned. It requires careful reading and revisits the concept of a function.

Suggested Problems

#6, 9, 13, 16, 19, 21, 22

1.7 FITTING FORMULAS TO DATA

Optional; if covered, one or two classes.

Key Points

Fitting lines and curves to data. Using technology to handle data.

Ideas for the Class

In this text there are many many graphs of real data that are taken as the starting point for a mathematical interpretation or analysis. However, sooner or later students will very wisely ask where the graphs themselves come from. You can tell them that it is a long and interesting story at the intersection of the subjects of statistics and mathematical modeling, and you can suggest with section 1.7 one of the simplest techniques that is used, namely regression.

If you or your students do not have access to technology to carry out regression, you way want to go very lightly on this section. On the other hand, with technology available, do not simply use it blindly. It is very easy in this section to violate the "black box principle", that technology should never be used to answer a question that the student does not understand. At the very least students must be encouraged to always graph the best fitting curve gotten from their calculator along with the original data points to see that their answer makes sense. And once again, warnings about the danger of extrapolation are essential.

You will probably have to address the technology directly. You may have to explain how to get the data points into the machine and how to get the numerical and/or graphical results out of the machine. It is a temptation to simply tell the student to get a manual and learn how to do it, but seeing the instructor do it validates it for the student.

If you wish to spend two days on this section, you may want to explain the least squares criterion. Begin with the problem of trying to fit a line to non-collinear data points. You might start with three points and choose a line through two of them and ask how far off the third point is (in the y-direction). Then choose another line and ask how far off the omitted points are, and so on. Get them used to the idea of using the distances from the missed points as a measure of the accuracy of the line, and have them suggest which of various lines might be a better fit based on distances they have computed. Of course the problem is that some lines will come closer to one point and others will come closer to another. To compare the goodness of fit, they need one number for each line, not a whole set of distances. The number often used is the sum of the squares of the distances, and the students can compute this for their lines. Tell them that the computer finds the equation for the line for which the sum of squares of distances is the least, called the *least squares line*. They and you will be happy to let the technology carry out the details.

Problems

A few of these problems (2, 5 & 10) do not ask the student to find regression curves, but to reason about them. The other problems are a mixture of linear and exponential regression. Once the technology has been mastered, the emphasis should on the interpretation, rather than on the calculations.

Suggested Problems

#1, 2, 3, 8

CHAPTER TWO

Six classes.

Overview

This chapter presents the concept of the derivative as a limit of difference quotients, focusing on the interpretations of the derivative as rate of change and as slope of a curve. The emphasis is on understanding where the derivative comes from, what it means, and how it can be applied. A concerted effort is made throughout the chapter to give graphical, numerical and verbal considerations equal billing. Since derivatives will be computed approximately by difference quotients or by estimating slopes, students can work with functions presented verbally, by tables of values, and by graphs as well as those defined by formulas. Because so much emphasis has traditionally been given to derivatives as formulas, it may seem that algebraic considerations are being given short shrift; wait until Chapter 5 where they are given full play.

2.1 INSTANTANEOUS RATE OF CHANGE

One class.

Key Points

Instantaneous rate of change of a function at a point as the limit of average rates of change over shorter and shorter intervals. The slope of the tangent line. Estimating rate of change with difference quotients.

Ideas for the Class

Using the idea of an object thrown straight up into the air, in this case a grapefruit, the text asks students to compute average velocities, studied in Chapter 1, over various portions of the trajectory and to reflect on the difference between each of these average velocities and the velocity at a point in time, which has not previously been discussed. You might characterize the average velocity of a car, for example, as the distance traveled divided by the time spent and the instantaneous velocity as the speedometer reading. Then ask students to speculate on what the speedometer reading might be in a car that traveled 30 miles in one hour (no way to tell), one mile in two minutes (not much better able to tell, but likely close to 30 mph some of the time), or 44 feet in one second (very close to 30 mph).

The key to instantaneous velocity is computing average velocities (with difference quotients) over smaller and smaller time intervals. On the one hand, students must see the need for going to average velocities over an interval to estimate the velocity at an instant. This can be explained visually, if you wish, using some pictures of animals in motion taken by Edward Muybridge [*Animals in Motion*, Dover, 1957] over a century ago. As you point to a particular frame and ask what the average velocity is there, it quickly becomes apparent that you need two frames to answer the question, and it is natural to select adjacent frames. On the other hand, students must appreciate the need to go to shorter and shorter intervals to improve accuracy.

The word limit is introduced but not formally defined. This is usually not a problem for students in this course. You might tell them that when you say that the instantaneous rate of change is the limit of average rates of change, this means that the instantaneous rate can be approximated by an average rate. It is vital though that students bear in mind that difference quotient approximations are not exact. It is also good for them to understand that in practical situations often an approximation is as good as you can do because the data they have to work with is itself not exact and may be incomplete as well - for instance table values are rounded off and skip many values. Fortunately a close approximation is often adequate for purposes at hand.

As you move to rates of change other than velocity, it is important to have students computing average rates, interpreting signs and other outcomes in practical terms.

Be sure to include some graphical examples. The slope of a curve at a point is well illustrated by having the student zoom in on a smooth curve until it appears straight. Compute some slopes of curves this way, and make sure the student makes the connection between average rates of change and slopes of secant lines on the one hand, and instantaneous rate of change and slopes of tangent lines on the other.

Ask the students to fall asleep for the next three nights reciting the words "approximated by difference quotients" to themselves one hundred times.

Problems

Problems 1-10 are relatively simple problems to reinforce the numerical and graphical interpretations of the derivative at a point. It is a good idea to work through a few of these in class before assigning others.

Problems 11-15 are more substantial and it would be wise, not only to assign most of these, but to discuss them as well. The concept of instantaneous rate of change is so fundamental that every effort should be made to ensure that students comprehend this fully.

Suggested Problems

#1, 2, 4, 5, 6, 7, 9, 10, 13, 14

2.2 THE DERIVATIVE

One class.

Key Points

Definition of derivative at a point as an instantaneous rate of change. Graphical and numerical computations.

Ideas for the Class

The derivative of a function at a point is defined as the rate of change of the function at that point. Having seen in Section 2.1 many examples of real world functions, the students are asked to work in this section with some 'pure' functions defined by formulas, graphs, or tables, but not arising from an applied context. The methods available to them for computing derivatives at this point in the course, though, are just the same.

If a function is defined by a formula or a table of values, students must approximate with a difference quotient over a short interval (or in the case of a table as short an interval as they have data for). If the function is defined by a graph they can compute the slope of a tangent line starting either by drawing a tangent line on the graph with straightedge or by zooming in on a calculator till the portion of the graph they are seeing appears straight.

Once again, emphasize the distinction between the derivative itself and its approximations by difference quotients.

Problems

These problems include graphical interpretations (1, 3, 4, 7, 8, 9, 10, 11 & 12) and numerical estimations (2, 5, 6, 13 & 14). Students might be more comfortable with the graphical interpretations than with the numerical estimations. Problems such as Problems 2, 5 & 7 should be handled with care. Time spent on these now will pay off later. Beware the difference in scales on the two axes in Problem 8. Problem 9 is a good problem for discussion in groups. Remind students that on the algebraic problems, they are not to use any short-cuts for the time being; but give them the option of checking their work with formulas if they can and wish to.

Suggested Problems

#2, 3, 5, 6, 8, 9

2.3 THE DERIVATIVE FUNCTION

One class.

Key Points

Understanding the derivative as a function. Finding the derivative function graphically and numerically. The derivative and increasing and decreasing functions.

Ideas for the Class

Point out that the derivative of a function at each point defines a new function, the derivative function. This is best illutrated graphically, and you may want to spend most of your class on graphical differentiation. Start with a smooth curve with some ups and downs on the board, and go through the details of sketching the derivative. Students will need lots of practice; the first six problems are good ones to have students work through in groups. In case the graph is given on a grid, as in Problems 1–6, students can estimate slopes by imagining tangent lines and then plotting points representing the slopes. Some easy points to plot first are the points where the derivative is zero and thus touches the x-axis. Point out, or have students notice, that the derivative is positive where the function is increasing and negative where it is decreasing, and thus is zero where the function changes direction. You will also need to reassure students that graphical methods, especially when we have no scaling on axes, are imprecise, but the global properties that emerge are consistent and your grading will be done on a consistency basis.

Students will start out examining graphs point by point, but move them toward global properties. You might sketch the graphs of the arctangent function and its derivative without identifying them, and ask which graph is the derivative of the other. The point is that one of the graphs is always increasing and that the other is always positive. You might sketch something like a sine curve as the derivative of an unidentified function and ask what it tells us about the original function. Then sketch the same curve shifted upward so that it never crosses the x-axis and ask the same question. Whatever else is said, your students should observe that since the derivative is always positive the original function must have been increasing.

An activity that produces a lot of mathematical discussion is to have each student sketch the graph of a function on a piece of paper, put his or her name on it, and pass the paper to a neighbor. The neighbor draws the graph of the derivative of the function on another piece of paper and copies the name of the first person on it. The original is passed back to its author and the derivative is passed to a third student. The third student sketches the graph of the original function from the derivative and then compares it with the original. This activity can be used for review later; even at the end of the term, many students may find it difficult.

Invent similar activites with tables of values. You want students to look for global properties, such as intervals where the function is increasing or decreasing.

Since it takes some time for students to come to terms with the graph of the derivative function, resist the temptation to make too many connections for them at first. For example, unless someone in class points out the relationship between zeros of the derivative and extrema of the original function, or between extrema of the derivative and inflection points of the original, don't do it for them. If the issues do arise, just note them as perceptive observations that merit future consideration and move on.

Note that finding formulas for the derivatives of functions given by formulas is deferred till Chapter 5. There is one example in the text, $f(x) = x^2$, that can be used to suggest the existence of formulas if you wish, or it can be omitted. For the sake of students who already know the algebraic shortcuts, point out that the way you show the shortcuts work is to use the definition of the derivative, starting with difference quotients. Some students have had the shortcuts hammered into them so thoroughly that they believe such shortcuts constitute all of calculus.

Problems

Problems 1-7 and 17-28 give practise in sketching the derivative function from a graph. Problems 8, 12-14, 29 and 30 give practise in reading properties of the derivative from a graph or in constructing a graph, given properties of the derivative. Problems 9 & 15 allow reasoning about the derivative from numerical data. Problems 10, 11 & 16 refer to algebraic derivation. All should be represented.

Motivate the students to look for qualitative, global features rather than detail.

Suggested Problems

#1, 2, 5, 12, 15, 16, 29

2.4 INTERPRETATIONS OF THE DERIVATIVE

One class.

Key Points

Using the difference quotient and units to discover the meaning of the derivative of real world functions. The dy/dx notation.

Ideas for the Class

This is one of the most important sections of the text! What is the meaning of a derivative in a real world situation?

The mathematical content of this section is not increased from that of the previous section. But the emphasis here is different, on using the difference quotient and units to give meaning to the derivative, especially as applied to rates. Students will get more out of a discussion of this material after they have wrestled with some of the homework problems. You may want to go over some of the examples and assign problems, then spend a class discussing their difficulties with the homework. Make sure you have overcome you own difficulties with the problems first!

Students find some of these problems hard. For example, Problem 13 discusses a function $g(v)$ giving the fuel efficiency in miles per gallon of a car going v miles per hour, and asks for the practical meaning of $g'(55) = -0.54$. One thing we want them to seize upon immediately is the negative sign but many students are so overwhelmed by the presence of the numbers and the unfamiliar combination of units that they never notice it. (We want them to say that the efficiency is decreasing by 0.54 mpg per mph at the speed of 55mph.) They usually understand after the discussion, but coming up with their own explanation can be a challenge.

Students may find some of the history of the derivative notation interesting. Newton's *dot* notation [$x(t)$ for $x'(t)$], introduced in the 1660s, is not now widely used. Leibniz's *double-d* notation [dx/dt for $x'(t)$], introduced in the 1670s, is widely used. The *prime* notation $x'(t)$ was introduced by Lagrange over a century later. Emphasize that in dx/dt, the dx is to remind us of a small change in x, and the dt is to remind us of a small change in t. Thus dx/dt is to remind us of the difference quotient which approximates the derivative and from which the derivative comes as a limit. To think of dx/dt as a fraction with products is wrong; canceling the d's would be akin to canceling the 2's in the fraction 27/28.

Problems

It is essential for students to verbalize as well as write down answers to problems in this section. Be picky as to how they answer the questions. When students can articulate their answers to this type of questions, they are well on their way to understanding the concept of the derivative as a rate of change.

A variety of interesting problems are presented in this section. It might be a good idea to do one or two together so that they realize how important the correct interpretation, correctly articulated, is.

Suggested Problems

#1, 3, 4, 10, 12, 14, 18

2.5 THE SECOND DERIVATIVE

One half to one class.

Key Points

Interpretation of the second derivative in terms of concavity. The second derivative as a rate of change.

Ideas for the Class

The most important thing for students to keep in mind about second derivatives is that they are derivatives in their own right and hence always measure a rate of change, namely the rate of change of the first derivative. The second derivative thus tells how the first derivative is changing, and for this to make any sense the meaning of the first derivative must be very clear. For instance, in a graphical situation the first derivative is a slope and so the second derivative tells how the slope is changing. In a distance-time situation, the first derivative is a velocity and so the second derivative tells how the velocity is changing. You might ask students in class to name some quantities that change over time, and in each case have them state what the first derivative means (as in the preceding section of the text), and then ask them what it would mean if the second derivative were positive or what it would mean if the second derivative were negative.

Go through the graphical interpretation of the second derivative in terms of concavity very carefully. The goal is for students to see the geometry so clearly in their minds that they are not even tempted to try to rely on memory to relate signs of first and second derivatives to shapes of graphs. Treat all four cases separately:

- increasing and concave up, like e^x, is increasing at an increasing rate;
- increasing and concave down, like $\sqrt{x}$, is increasing at a decreasing rate;

- decreasing and concave up, like e^{-x}, still has an increasing derivative, but the language is trickier;

- decreasing and concave down, like $-e^x$, has a decreasing derivative.

It is easy for the students to confuse *increasing first derivative* with *first derivative of increasing magnitude*. If your examples have not cleared things up, point out that the sequence $\{-1, -2, -3, \ldots\}$ is decreasing, while the sequence $\{-1/2, -1/3, -1/4, \ldots\}$ is increasing. Determine whether they understand by having them write out, in clear English, what the second derivative tells them about concavity, and an explanation of why in terms of changing slope.

Acceleration has not been explicitly mentioned in this section or in the exercises. If you feel that the concept is important for your students, it is not difficult to invent some simple examples that make the idea clear.

Problems

Most of these problems require interpretations of the sign of the second derivative. Only Problem 10 requires the calculation of a second derivative. Graphical interpretations (2-4, 11, 13-14) are the most prevalent. These should be well represented as assigned problems. Problems on describing the behavior of the second derivative from a table of data (5 & 12) should also be assigned. Problems based on quotes (6-8) are good discussion and group work problems.

Suggested Problems

#1, 2, 4, 5, 8, 12

2.6 MARGINAL COST AND REVENUE

One class.

Key Points

Derivatives and the concept of marginality. Marginal cost and marginal revenue. Maximizing profit.

Ideas for the Class

A discussion of what a cost curve means, as well as interpretations of its slope and concavity, is a good place to start. How is economy of scale reflected in the shape of the graph? Follow with a discussion of revenue curves, both the straight line (fixed price) case and the concave down (quantity discount) case. Then put cost and revenue together and discuss profit. Students sometimes forget about costs, wanting to maximize revenue rather than profit, so keep reminding them that revenue and costs must always be considered together. Note that in this section we consider only cost and revenue, not demand, supply, or other things, so that our analysis is not entirely realistic.

It is easy to memorize without understanding the fact that marginal cost of 100 items is the approximate cost of producing the 100th item. Be sure your students understand why this is true. You might ask them to explain it clearly in writing using a difference quotient approximation for the derivative of the cost function.

When discussing profit, explain why (or have students explain why) increasing production increases profit when marginal revenue exceeds marginal cost, and why decreasing production increases profit when marginal cost exceeds marginal revenue. (Assume revenue exceeds cost.) Use the tangent lines to conclude that when profit is maximized, $R'(q) = C'(q)$. This is an entirely graphical explanation, but it is both clear and convincing to the students.

Problems

Problems involving graphs (1-4, 7, 8, 13-15) form the backbone of this set of problems. Problems involving tables of data (5, 6 & 9) should also be assigned. Problems requiring written or verbal interpretations such as 10-13 are essential for developing verbalization and written thoughts.

Suggested Problems

#1, 4, 5, 6, 7, 9, 10, 15

CHAPTER THREE

Six or seven classes.

Overview

The purpose of this chapter is to give the student a practical understanding of the definite integral as a limit of Riemann sums, and to bring out the connection between the derivative and the definite integral in the Fundamental Theorem of Calculus. The motivating problem is to compute the total distance traveled from the velocity function. We use the same method as in Chapter 2, introducing the concept graphically and numerically without going into analytical techniques. Definite integrals will be computed approximately by Riemann sums or by estimating areas, and so students can work throughout with functions presented by tables of values and by graphs as well as those defined by formulas. The student should finish the chapter with the ability to estimate a definite integral and an understanding of how to interpret it as area under a curve, average value of a function over an interval, and most importantly, as total change in a quantity given its rate of change.

3.1 ACCUMULATED CHANGE

One class.

Key Points

Approximating the distance traveled, given the velocity function, by using the formula

$$\text{Distance} = \text{Rate} \times \text{Time}$$

to approximate the distances traveled over short time intervals. This leads to the idea of a Riemann sum. Graphical interpretation (area under the velocity graph). Extension to rates and changes other than velocity and distance.

Ideas for the Class

This section is your chance to help the students see that a Riemann sum is not anything mysterious, that it is a natural generalization of the formula Distance = Velocity × Time with which they have long been comfortable. To help them discover these sums on their own, the text begins with a very simple problem: I traveled 30 miles/hour for 2 hours, then 40 miles/hour for 1/2 hour, then 20 miles/hour for 4 hours. How far did I go? Instantly the students can figure this out, and so get

$$\text{Distance} = \text{velocity} \times \text{time} + \text{velocity} \times \text{time} + \text{velocity} \times \text{time}$$

It is clear why sums are necessary—because velocity changes over time! And is clear why the problem is easy—because the velocity is not changing all the time!

Students at this point are just one step away from a full Riemann sum, which will come out of considering the thought experiment in the text, How Far Did the Car Go?, in which velocity does change all the time. The big difference is that they will be forced to accept an approximate rather than an exact distance for the answer. But then they have been living with approximations ever since they learned of instantaneous rates of change. You can tell them that the need for approximation is what makes the subject calculus rather than algebra.

Go over the thought experiment in the text, but using different numbers. Better yet, have students tackle a similar problem before you talk about it, any of Problems 1, 2, or 3 is appropriate. For a more colorful problem you might like to have students get into groups of four or so and try a problem like the following:

Example 1   Jan and Pat are driving along a country road at 45 miles per hour. As the car rounds a curve, Jan sees a skunk in the middle of the road about 100 feet ahead. Jan immediately applies the brakes, and Pat notices that the speed of the car for the next three seconds is as given in Table 3.1.1:

TABLE 3.1.1

time (sec)	0	1	2	3
speed (mph)	45	35	23	0

Does the car hit the skunk?

Start with a table of velocities; don't use any graphs at first. Have students give upper and lower bounds for the distance traveled, using left and right hand sums. If they suggest averaging the two for an improved answer let them know that that their idea is a good one. Demonstrate that more data (sampling velocity more often) will make the upper and lower bounds closer together. As much as possible, get the students to give the estimates. Then have the students translate the table and the estimates into graphical form. Afterward, group discussion can clear up remaining questions; often students themselves make the explanations you plan to give.

When you make a graph of velocity vs. time, be prepared for confusion. Students are accustomed to graphing distance vs. time and are likely to want to interpret this graph the same way. They may be confused by the fact that an area under a graph can represent a distance. While they may be used to letting a length represent a variety of quantities, they don't have much experience letting an area represent anything but area. It may help to remind them that on the graph, horizontal lengths represent time and vertical lengths represent velocity.

For another approach to the ideas of this section, couched in a more general rate of change setting, you might use the following problem instead of a velocity problem.

Example 2   Land management officials notice that an introduced species of tree is making serious inroads into an ecosystem. (An example is the tamarisk tree in the Great Basin of the western United States). In a certain area, the number of new trees per year is increasing every year. Some of the growth rates are given in Table 3.1.2.

TABLE 3.1.2   *Rate at which new trees are appearing*

Year	1950	1960	1970	1980	1990
Trees/year	337	371	408	448	493

We want to estimate the total number of new trees that have appeared between 1950 and 1990. Answering these questions will show you how such an estimate can be made.
(a)   What is the minimum number of new trees that could have appeared between 1950 and 1960? The maximum number?
(b)   What is the minimum number of new trees that could have appeared between 1960 and 1970? The maximum number?

(c) During these four decades, from 1950 to 1990, what is the minimum number of new trees that could have appeared? The maximum number? Did you use the assumption that the number of new trees each year is increasing? Where?

(d) If you had to guess how many new trees appeared between 1950 and 1990, what would be your guess? What is the maximum possible error in your guess? (That is, what is the maximum possible difference between your guess and the true value?)

Suppose that some additional growth figures are obtained, and presented in Table 3.1.3.

TABLE 3.1.3   *Rate at which new trees are appearing*

Year	1955	1965	1975	1985
Trees/year	343	389	418	467

(e) Is the second set of information consistent with the first? In what interval would the number of new trees in 1955 have to be in order to be consistent?

(f) Recalculate the left and right hand sums in the light of this new information. Make a new guess for the total number of new trees that appeared between 1950 and 1990, and estimate the maximum possible error in your guess.

(g) If the numbers of new trees were known for each even-numbered year, by how much would your left and right hand sums differ? What if the numbers for each year were known?

(h) How accurately do you think the total number of new trees may be calculated in this example? Why?

Problems

Problems 1, 2, 3 & 6 are all based on tables of data. Note that in Problem 2 the speed is given in *mph* and time is given in *min*. Most students will not notice this at first. Problems 7 & 9 are based on graphs and students may need some guidance on these. In Problems 5 & 8 lies a temptation of using formulas for students who may have had calculus before. Such students may want to use the Fundamental Theorem. Sketching and explaining are required which should steer them in the right direction.

Suggested Problems

#1, 3, 5, 7

3.2 THE DEFINITE INTEGRAL

One or two classes.

Key Points

The definite integral of a function as a limit of Riemann sums. Estimation of an integral using left and right hand sums. Evaluation with calculator. Interpretation as total change and as area.

Ideas for the Class

Problems in Section 3.1 have led to calculations of the form

$$\text{Distance} \approx \text{velocity} \times \text{time} + \text{velocity} \times \text{time} + \text{velocity} \times \text{time}$$

and

$$\text{Area} \approx \text{height} \times \text{width} + \text{height} \times \text{width} + \text{height} \times \text{width}$$

which, after a brief reminder, you can write one after the other on the board. You could talk about different problem situations whose solutions lead to the same computations, and tell your students that it is a mathematician's job to keep an eye out for such situations so that the underlying computational scheme, which is likely to be useful, can be studied carefully.

In this case we define the definite integral. The word limit is used in the definition, but not itself formally defined. Students understand the need for smaller and smaller increments in the Riemann sums which correspondingly have more and more terms. Just make sure they understand that when you say that the definite integral is the limit of Riemann sums, this means that the definite integral can be approximated by Riemann sums.

Start off by computing a few left and right-hand sums by hand. For example, estimate $\int_0^1 x^2 dx$ by making a table of values for $x = 0.0, 0.1, 0.2, \ldots, 1.0$, adding up all the values but the last one, and multiplying by $\Delta x = 0.1$. Do the similar thing for the right hand sum. Better yet, have the students do it, perhaps in groups. For a good estimate, average the left and right-hand sums. Students should have this approach down cold before turning it over to the technology. Do another example by hand in which $\Delta x \neq 0.1$. You might try $\int_1^3 \sqrt{x}dx$ with $n = 3$. Common mistakes are to always multiply by 0.1, or to forget to multiply by Δx altogether.

You will have to decide whether to introduce the Sigma notation in your class; it is not necessary to do so, but even if you do, don't rush to get to it. Students may have trouble reading an expression such as

$$f(t_0)\Delta t + f(t_1)\Delta t + f(t_2)\Delta t + \ldots f(t_{n-1})\Delta t$$

which you may want to spend some time with first. Make sure that the connection between whatever symbolism you introduce and the thought experiment is as obvious as possible; draw lots of graphs.

Do not spend much time in this section on the interpretations of the definite integral as total change and area because you will have ample opportunity to do so in the rest of the chapter.

After computing several Riemann sums, your students will be delighted to evaluate definite integrals for functions given by formulas using the integration package in their calculator or computer. You may have to show them how to do it, though.

Emphasize that the definite integral is a number. Some students who have had calculus before may think it is an antiderivative; others think that $\int_a^b f(x)dx$ is *defined* to be $F(b) - F(a)$. They may want to know why you aren't doing it "the easy way" using the Fundamental Theorem, If they do ask, you can point out that graphical and tabular examples can not be done that way, nor can seemingly simple integrals such as $\int_0^1 2^{x^2}$ (the integrand does not have an elementary antiderivative).

Finally, once again, emphasize the distinction between the definite integral itself and its approximations by Riemann sums. Riemann sum approximations are not exact, but often an approximation is as good as you can do because the data you have to work with is itself not exact and may be incomplete as well. Fortunately a close approximation is often good enough for purposes at hand.

Ask the students to fall asleep for the next three nights reciting the words "approximated by Riemann sums" to themselves one hundred times.

Problems

Many of these problems require technology. You may want to discuss the fact that accuracy to one decimal place means that the error is less than 0.05. Do not underestimate problems 1-12. It is valuable for students to do a number of these in order to gain confidence in dealing with Riemann sums. Problems 13 & 16 provide the opportunity to discuss functions that are not monotonic over the given intervals. Discuss splitting the interval so that the function is monotonic over each subinterval. Problems 14 & 15 are good problems to assign because they deal with a table and a graph respectively. At least one of Problems 17-19 should be assigned and discussed. Problem 20 needs careful thinking and is a good group work problem.

Suggested Problems

#1, 11, 14, 15, 19

3.3 THE DEFINITE INTEGRAL AS AREA

One class.

Key Points

Basic interpretation of the definite integral as (signed) area.

Ideas for the Class

This section reinterprets the definite integral in terms of area. The discussion up to this point has involved only positive functions, and continues in that vein to start. Then the switch to functions that are not always positive is made. Make sure that the student distinguishes between the value of a definite integral (a number, which can be positive, negative, or zero) and an area (never a negative numer). Stress that area is only one interpretation of the definite integral.

Be sure to indicate how to approximate the definite integral of a function graphed on graph paper by counting the grid boxes under the graph and multiplying by the area of the individual boxes. This method is a very simple substitute for a Riemann sum that will be used several times later in the text.

Problems

All the problems in this section require of the student to use graphs and in many cases to make numerical estimates. The link between the sign of the integral and area should be the focus here. Doing estimations (counting squares) could require some guidance as students tend to feel somehow insecure when doing this. Make students aware of helpful symmetry assumptions.

Suggested Problems

#1, 2, 3, 6, 17

3.4 THE DEFINITE INTEGRAL AS AVERAGE VALUE

One half to one class.

Key Points

Average value of a function. Graphical interpretation of average value. Its expression and computation as a definite integral.

Ideas for the Class

Average value of a function should be explained both graphically and in terms of Riemann sums. Probably students will like the graphical explanation best. Ask them to imagine that the region under the graph of a function between high walls $x = a$ and $x = b$ is made of wax. Then the average value of the function between a and b is the level the wax would reach if it were melted and allowed to flow under the influence of gravity, but always staying between the walls. Give some graphical examples in which students are asked to guess the average value from the graph before calculating it numerically. They enjoy getting picky, trying to place a horizontal line on the graph so that the area above the line but beneath the function exactly matches the area below the line but above the function.

Students who become confused with average values are usually those who have had calculus before and think of the definite integral strictly in terms of an antiderivative.

Problems

Problems 1, 2, 5 & 12 ask of the student to deal with average value, strictly graphically. Problem 9 refers to the absolute value of f—illustrate this notion with an example function such as the absolute value of $f(x) = x^2 - 1$. Problem 12 is a valuable problem—the notions of derivative and integral are dealt with in the same problem. A problem such as Problem 2 links the intuitive visual notion of average value and the formula for average value. Be sure to assign this problem. Problems 3, 4 & 6 are simple forerunners of more substantial problems such as 7, 8 & 11. Refresh their understanding of exponential growth and decay by comparing Problems 8 & 11 (and also involve 7).

Suggested Problems

#1, 2, 4, 7, 11

3.5 INTERPRETATIONS OF THE DEFINITE INTEGRAL

One to two classes.

Key Points

Using units and the interpretation as total change to discover the meaning of the definite integral of real world functions.

Ideas for the Class

Thinking of the definite integral as the limit of Riemann sums allows students to figure out units for the integral, and these units are invaluable for interpreting the integral. In particular, look at examples in which the integrand is a rate of change (velocity, growth rate, marginal cost/revenue). It is essential that students understand that the definite integral of a rate of change is the net change.

Students are sometimes amused when they compare the area interpretation of the definite integral with units in a graphical situation. Ask, for example, what are the units of area in Problem 9? Thinking of area as height times width, and taking units on the axes for height and width, gives area in gallons/hour × hours = gallons. The units of area are gallons and one grid box on the graph has an 'area' of 1/16 gallon! At this point students will easily compute the integral and get the units right just by counting grid boxes.

Example 6, on the changing carbon dioxide content of pondwater over the course of day, is a serious and extended ecological application of just about everything that the students have learned about the definite integral. You may want to assign it for reading, then take a good bit of time with it in class, especially if you have life sciences students in your course.

Problems

Problems 1-7, 14 & 16 concern the interpretation of the integral as a rate of change. Problems 9, 12, 13, and 17 - 19 do the same thing graphically. Problems 8 & 14 do it numerically. Represent each of these groups in the set of assigned questions. The interpretation of the integral as a rate of change should, as with the interpretation of the derivative, be verbalized and written down. Be very picky. Graphical interpretations are confusing to the students at first and they might need more guidance here. The numerical interpretations should serve to refresh Riemann-sum calculations.

Suggested Problems

#1, 2, 6, 9, 10, 12, 15

3.6 THE FUNDAMENTAL THEOREM OF CALCULUS

One class.

Key Points

The Fundamental Theorem of Calculus, explained by considering the definite integral of a rate of change as total change.

Ideas for the Class

Students saw in Section 3.5 many examples of definite integrals of real world rates of change and their interpretations as total (net) change. In this section they will work with 'pure' functions defined by formulas, graphs, or tables, but not arising from an applied context. It is not a big step for them. The rate of change of a physical quantity is merely replaced by the derivative of some function. The Fundamental Theorem allows them to do this, asserting that the definite integral from a to b of the derivative of a function equals the change in the function between a and b.

Once the theorem is stated, the students can begin to apply it, and the applications are primarily graphical and numerical. Since analytic antidifferentiation has not been developed, most of the applications will consist of using the definite integral (computed by Riemann sums or estimating areas) to deduce values of a function given its derivative, as in Examples 3, 4, and 5. In Example 4, for instance, students will read from the graph that F' is positive for $0 < x < 50$, so they will know that the function F itself is increasing over this range. They will need the Fundamental Theorem to answer the natural question, 'How high does it go?'

Problems

Problems 1 & 3-5 emphasize the relationship between a function and its rate of change. The notion of a function value of F as an area below the curve of F' takes some getting used to. So tread lightly here. The two different approaches—area interpretations and numerical estimations—should be handled with care. Problems 7 & 8 are straightforward applications of the Fundamental Theorem. The functions involved are simple and no difficulties should be experienced here. Problems 9 - 14 involve graphical interpretations and could be useful for group work.

Suggested Problems

#1, 3, 4, 5, 7

CHAPTER FOUR

Six to eight classes.

Overview

This chapter deals with a few of the families of functions that arise most frequently in mathematical modeling. We begin by returning to the exponential family, expressing it in terms of the base e as is customary in applied disciplines. The natural logarithm is introduced, which will facilitate work with exponential models. Included as well are polynomials, sine, and cosine. The last two sections, on the logistic family and its use in population models, and on the surge family and its application to drug concentration in the blood, can be viewed as serious extended examples of the modeling process.

4.1   THE NUMBER e

One class.

Key Points

The number e. The representations a^t and e^{rt} for the exponential family. Continuous growth and decay rates. Modeling growth and decay, including continuously compounded interest and effective annual yield.

Ideas for the Class

The main thing you want the students to understand is that every exponential function can be expressed in two ways, either as a^t or as e^{rt}. There is no difference in the graphs or tables of values or applications. These are just two ways of saying exactly the same thing. The parameter a is called the growth factor, and the parameter r is called the continuous growth rate. If $a^t = e^{rt}$ then $a = e^r$ and so for r very near zero $a \approx 1 + r$, but this is not even close for large values of r.

It is a good idea to have students express $e^{0.2t}$ in the form a^t and graph and evaluate both functions. And then go the other way, starting with $(1.03)^t$ and finding its expression as e^{rt}. (Since logarithms have not yet been studied, students will have to solve for r graphically or by trial and error, but that should not present a problem.) Repeat with a decaying exponential.

Be sure to leave time for some growth and decay problems, since they are the reason the students are studying exponentials. Students will have to solve graphically for times, doubling times, half lives, etc., which may help them to appreciate logarithms in the next section. If you elect to teach continuously compounded interest, note that the effective annual yield of an investment of value $a^t = e^{rt}$, where t is in years, is defined to be $a - 1$, which is the growth rate over one time period already studied in Section 1.5 (*not* the same as the continuous growth rate r). Try not to get bogged down with the terminology.

An example for the class to think through is the following:

Example 1   Proud parents of a newborn child establish an educational fund for the child by depositing $2000 in an account that pays 6.25% interest, compounded continuously.

(a)   How long will it take for the account to double in size? (Use trial and error.)

(b)   What will the amount in the account be in 18 years?

(c)   If the parents want the account to amount to $20,000 after 18 years, how much should they deposit initially?

Your students already know exponential functions from Section 1.5. Strictly speaking, there is no mathematical need to write all exponentials with the same base e. There are in fact many situations, such as the Mexican population example in Section 1.5, where taking the annual growth rate as base is very natural. So students may ask, why this section? For now it is probably best to appeal to convention. In the applied literature from science to engineering to economics the exponentials that arise are usually expressed in the form e^{rt}. The calculus-based reason for this is presented in Section 5.2.

Problems

Most of these problems are straightforward but require careful reading. The first few problems simply emphasize the difference between annually compounded versus continually compounded interest. Graphical interpretations such as Problems 5 & 10 are necessary for visual comprehension. Problems 12 & 13 revisit exponential decaying functions. Be sure to assign at least one of these. The function in Problems 15 and 20 is, of course, the one used in normal distributions. Although they might not have dealt with normal distributions, they should be familiar with the idea. Use this to initiate a discussion. Also point to the fact that the Fundamental Theorem is useless for Problem 20. Problem 16 is a forerunner of logistic growth. Time spent here in investigating this problem is sure to pay off later. Problem 22 is difficult because of the large numbers involved, but calculators can handle them easily and students like it. (By the way, in February 1992 the Supreme Court refused to hear the De Haven case, saying that because of the Statute of Limitations, the viability of the case had expired 123 years earlier!)

Suggested Problems

#1, 9, 10, 13, 14, 16, 21

4.2   THE NATURAL LOGARITHM

One to two classes.

Key Points

Natural logarithms and their use in solving exponential equations. Half lives and doubling times. Relationship between a^t and e^{rt}.

Ideas for the Class

It is easy to state the point of this section. In problems involving exponential growth and decay, time appears as an exponent. If the problem asks for the time when something happens, the easiest way to solve for it is to use logarithms.

In this text only the natural logarithm is used. Thus $y = \ln x$ if and only if $e^y = x$. Focus on the idea that the logarithm is an exponent. Your students may need some practice with the basic rules for computing with natural logarithms, and you might even like to give some indication of why they hold. But spend as much time as you can with the applications examples and problems. Students do not find them easy, but they find them interesting and will enjoy a classroom discussion after they have been given a chance to struggle with them overnight on their own.

Problems

It is essential for students to develop a "feel" for logarithms. For this it is necessary to select a number of the routine problems (4-18, 33-41) before answering the many "How long?" and "When?" questions. Also point out how the ln-function takes giant numbers and reduces them to midgets as happens in Problem 30. In both Problems 44 & 46 it is not necessary to have a number value for the initial quantity in order to answer the question. This is a useful principal and might need a bit of explaining. Problem 49 is significant because it links logarithms with the main theme of the book. Use this to generate a general discussion on the nature of the derivative of the logarithmic function. Problem 50, on the other hand, links the logarithmic function with integration and so the integral of the ln-function, as such, should be discussed before tackling this problem.

Suggested Problems

#4, 8, 16, 17, 19, 25, 26, 29, 30, 34, 35, 38, 45, 49

4.3 NEW FUNCTIONS FROM OLD; POLYNOMIALS

One class.

Key Points

Shifting and stretching graphs of functions; Sums of functions. Graphical properties of polynomial functions; Resemblance to power functions on large scale.

Ideas for the Class

The theme for the day can be families of functions; every function $f(x)$ fits into several families. Your students should become comfortable with: $f(x) + k$, $f(x - h)$, and $Af(x)$, the families of vertical shifts, horizontal shifts, and vertical stretches/compressions (with flip if $A < 0$) of f. Students can demonstrate the meaning of the parameters h, k, and A by sketching lots of graphs. The function $f(x) = x^2$ is a good starting place for the shifts, and $f(x) = x^3 - x$ shows up the stretches very nicely. The effects of the shift and stretch transformations are also easily demonstrated numerically by creating tables of values. Tables are especially helpful in explaining the direction of a horizontal shift, which students often find confusing. Algebraic work can consist of starting with a function given by a formula and a graph, then shifting and stretching to produce a new graph, and finally getting the formula for the new graph. This also provides a good review of function notation.

Polynomials can be motivated by the desire for functions whose graphs have both ups and downs. After some random experimentation with calculator or computer to suggest the variety of graphs possible with polynomials, you want the students to develop a bit of intuition. The degree of the polynomial limits the number of peaks and valleys. The global shape is the same as that of the leading term. Very near $x = 0$ the constant and linear terms give the right picture. If asked to graph $y = x^3 - 10x + 50$, for example, students should draw a negatively sloped piece crossing the x-axis at $y = 50$ that must be connected to distant arms in the first and third quadrants, and the degree limits the ways the connection can be made.

The significance of leading terms is brought home dramatically by graphing in large windows. Have students plot the two functions $y = x^3 - x + 83$ and $y = x^3$, first in a window $[-10, 10]$ by $[-100, 100]$ and then in a window $[-50, -50]$ by $[-10,000, 10,000]$. Then give them two other polynomials with the same leading terms and have them choose two windows, one where the polynomials have totally different graphs and one in which they appear to coincide. (Writing the polynomial so the terms are not in the usual order helps to get the point across.)

Relate the coefficients to the shapes. Ask questions like, "How do you get an upside-down parabola? What if you want it to go through the origin, too?" Relate the leading coefficient to the idea of vertical stretch or compression, and illustrate numerically. You could point out, for example, that if the leading coefficient is large then only a small value of x is needed to give large values of y so the graph is going to be narrow.

You might give some examples requiring students to select coefficients to fit certain criteria. For example, given a parabolic mirror 2 m wide and 5 cm deep at the center, write down a quadratic polynomial whose graph is a cross-section of the mirror. This is a good problem because students must also set up the coodinate axes.

Problems

The large variety of problems in this section range from simple shifts and stretches (1-12, 14), to polynomial investigation (16-24) and to more substantial problems where interpretation and estimation are prominent (13, 24-31). The emphasis should be on the latter group of problems, but it is necessary to start with the first group. Problems 13, 25 & 26 revisit supply and demand curves. One of these could be done in class and the others assigned for group work. These problems are good for generating an overall picture. Problems 28-30 need careful reading and careful thinking followed by a thorough discussion.

Suggested Problems

#1, 5, 6, 13, 15, 24

4.4 PERIODIC FUNCTIONS

Optional; if covered, one to two classes.

Key Points

Periodic phenomena. Amplitude and period. Sine and cosine functions.

Ideas for the Class

Note that this section is *not* titled "Trigonometry." Sine and cosine are used here only for modeling periodic phenomena. There are no angles, triangles, or circles mentioned in the text, and your students will not need to know how they relate to sine and cosine. Work in radians; make sure the students' calculators are set in radians.

Start by showing examples of periodic phenomena—cardiogram, annual climate data, oscillating time series, etc., similar to the examples in the book. Then show sine and cosine, the 'pure' oscillations. It will be clear from the examples that for mathematical modeling you will need to have available pure oscillations of all amplitudes and periods. In short, you will need a whole family of pure oscillations. Move to sketching graphs of $y = A \sin Bt$, or $y = A \cos Bt$ if you prefer. If students can do it for themselves on their calculators or on a computer, so much the better. Then introduce a vertical shift, a baseline different from 0, by considering $y = C + A \sin Bt$. Then show graphs of some of these, and have the students discover the amplitude and period and hence recover formulas for the functions.

Then come back to modeling. Example 6 makes for a great class discussion. If you plan to take a little time with this topic, Exercises 15, 16, and 17 are well worth the effort.

Students may ask about modeling the more complicated phenomena that are not pure sines. You might tell them that more complex periodic functions can be approximated by combining several simpler ones. Have them try graphing, for example, $y = 4 \sin 3x - 3 \cos 2x$ or $y = 4 \sin(x/2) \cos(5x)$. Encourage them to experiment, or announce a prize for the most interesting periodic graph on a calculator.

Problems

Students like problems on biological processes such as in Problems 2 & 3. Point out and discuss the word "approximately" in Problem 2 and discuss the fact that the graph in Problem 3 is not perfectly periodic. For Problems 4-9 the idea is to make conclusions from graphs generated by technology. It might be useful to draw their attenion to the fact that for a function $y = A \sin Bt$ the period is given by $2\pi/B$. Problem 11 is a good problem that can be applied in many situations. Make sure that students are on top of this problem, both graphically and algebraically. Follow this up by Problem 15 and discuss why the period-formula does not apply here. Problem 16 links the previous section (on shifts and stretches) with the current one.

Suggested Problems

#1, 3, 4, 7, 17, 18

4.5   THE LOGISTIC CURVE

One to two classes.

Key Points

The mathematical modeling process illustrated with the US population. The logistic function and its applications.

Ideas for the Class

Students are already familiar with the exponential growth population model. You might ask them how realistic they think it is, especially for the long term. Can the population of the earth grow forever? They may suggest that it will eventually level off, and so see that the exponential model has some limitations. This is demonstrated quite clearly in the text with actual US population data. In the period 1790–1860 an exponential model works extremely well, but after that rate of growth slows. A different model is needed.

The family $y = \frac{L}{1+Ce^{-kt}}$ of logistic functions models growth in an environment with limited carrying capacity. Students can gain familiarity with the family by graphing with different choices for the parameters L, k, and C, each one of which has a graphical significance. (L is the carrying capacity, k controls the rate of convergence to the carrying capacity, and C causes horizontal shifts in the time scale.) The fun is in the applications, where logistic functions are used not only for population studies but to track sales of new products (market eventually saturates) and physiological responses to doses of medicinal drugs (beyond a certain point more of a drug has no additional effect).

Problems

Problem 2 latches on to Problem 16 of Section 4.1. Doing both these problems makes sense and will make the students feel in control. Problem 1 could be assigned as a group project. Also ask students to draw up a model of the way news spreads within their own city. The numerical problems are Problems 3, 5 & 11 and should by now pose no threat to students. Similarly, students should enjoy the graphical interpretations (6, 7, 14, 15 & 16). Problems 10 & 12 should be emphasized for the fact that they carry the thread of the book through. Problems that have to be closely monitored are Problems 4 & 8 because of the variety of possible answers.

Suggested Problems

#2, 3, 4, 7, 8, 10, 14

4.6   THE SURGE FUNCTION

One class.

Key Points

The family of surge functions. Its use in modeling biological phenomena.

Ideas for the Class

The family $y = ate^{-bt}$ of surge functions has been chosen for the final section of this chapter because of its widespread use in modeling drug concentration in the blood over time. It will be a new family of functions for your students, and so you will want to begin by encouraging them to explore the graphical significance of the parameters a and b. But the bulk of the time should be spent with the models themselves and the data they model. The students should see that making the model amounts to finding suitable choices for the parameters. They may object that the graphs from the models do not exactly trace the curves of experimental data, but they need to know that it is in the nature of a mathematical model to make simplifying approximations. The art is to simplify enough to bring out the features of interest without simplifying so much that important information is lost. Finally, and perhaps most importantly, students should learn to read the drug concentration graphs for practical information, such as drug absorbtion properties and minimum effective concentrations.

Problems

These problems require very few calculations. The emphasis is on interpreting general trends. Students normally find these problems interesting and do not mind spending time on them. Students do, however, lean towards scanty explanations and interpretations - alert them to that. It might be wise to do a few of these (3, 6 & 11) in class in order to encourage well-motivated answers, assign others as group work (7 & 12) and a few of the remaining ones as assigned problems.

Suggested Problems

#1, 2, 5, 9, 10

CHAPTER FIVE

Six classes.

Overview

This chapter develops the basic differentiation formulas. Included are the standard combination rules (sum, difference, product) and the chain rule, as well as formulas for differentiating powers, polynomials, exponentials, logarithms, and the sine and cosine. Informal justifications are given, using graphical and numerical reasoning where appropriate. The final section touches on antiderivatives.

    The title of the chapter is intended to remind students that the formulas do not constitute the definition of the derivative; plenty of problems throughout the chapter review the concept and interpretations of the derivative.

5.1 DERIVATIVE FORMULAS FOR POWERS AND POLYNOMIALS

One class.

Key Points

Differentiation formulas for constant functions, linear functions, constant multiples, sums, differences, power functions, and polynomials. Using them in combination with graphical and verbal interpretations of the derivative.

Ideas for the Class

Students who have seen calculus before will be relieved to have come to the easy stuff (maybe you, too!). Don't forget that many are seeing it for the first time, however.

    The differentiation formulas introduced in this section are very easy. But it may take some time and effort for the students to connect them with the graphical, difference quotient, and verbal (rate of change) concepts of the derivative that they know from Chapter 2. Making this connection should be the main work of the section. To facilitate this process the algebraic techniques for proving the formulas have been postponed to Section 5.6.

    Use graphical arguments when you can, such as to predict the derivative of a linear function. It is usually worth asking what the graph of the derivative looks like (starting with the graph of the function), before you calculate the derivative. In a good many problems, maybe over half the ones you assign and discuss, ask for something more than the formula for a derivative. It is almost always a good idea to ask for an evaluation at a point, which can be done directly by asking for a comparison with a difference quotient approximation or indirectly by asking for the equation of a tangent line. And finally be sure to include some word problems in which the value of the derivative at a point must be both computed and interpreted in practical terms.

Problems

Problems 1-21 are routine practice of which students should do as many as it requires to fully master this rule. This should be followed by a selection of Problems 22-27 where specific derivatives are required. Higher derivatives form the theme of Problems 28-30 and tangent lines form the theme of Problems 31-33. The rest of the problems are application type of problems - some of which refer back to previous topics such as supply and demand (35, 38 & 42) and cost and revenue (34 & 39). Problems 44 & 45 are difficult and should be discussed.

Suggested Problems

#1, 3, 8, 21, 23, 26, 31, 34, 37, 39

5.2 EXPONENTIAL AND LOGARITHMIC FUNCTIONS

One class.

Key Points

Differentiation formulas for e^x, a^x, and $\ln x$. The significance of the number e.

Ideas for the Class

It is easy to see graphically that the derivative of an exponential function resembles the graph of an exponential function—for exponential growth both graphs start near zero and rise faster and faster as you move to the right. Using a graphical plausibility argument, the text gets to the equation $\frac{da^x}{dx} = c \cdot a^x$ for some constant c. A difference quotient approximation for $\frac{da^x}{dx}$ at the origin shows that $c \approx \frac{a^h - 1}{h}$ for h near 0. Now you and your students can have some fun with the mysterious (to them) number c. Have them help you make a table of values, say, for $\frac{a^{0.001} - 1}{0.001}$ as a function of a, each entry of which gives a new derivative formula of the form

$$\frac{da^x}{dx} \approx \frac{a^{0.001} - 1}{0.001} a^x$$

After a few computations have been done, casually ask what value of a would lead to the formula $\frac{da^x}{dx} = a^x$ and your students will find e for themselves. Then have them compare their values for c with $\ln a$. At last they will have a mathematical reason for interest in the base e.

For a class strong in algebra you might want to include the simple proof that

$$\lim_{h \to 0} \frac{a^{x+h} - a^x}{h} = \left(\lim_{h \to 0} \frac{a^h - 1}{h}\right) a^x$$

but in most cases it would be best to do this only for specific choices of a such as 2, 3, and 0.5. This is done for the base $a = e$ in Section 5.6 of the text.

Don't forget that the point of the derivative formulas for a^x and $\ln x$ is to use them! There are some applied examples and problems in this section that will make for good discussions.

Problems

The routine problems (1-20) are followed by other technical problems (21-22 & 24). These provide the necessary basis for tackling the applications (25-29). In an overview of these, let students refresh their knowledge on exponential growth and decay by pointing to what problem represents what. Also point to specific percentage increases or decreases. Problems 30-35 are of a somewhat difficult technical nature and should be reserved for students who seek further enrichment.

Suggested Problems

#1, 9, 14, 20, 22, 25, 29

5.3 THE CHAIN RULE

One class.

Key Points

Recognizing when the chain rule is necessary and how to apply it.

Ideas for the Class

The list of functions whose derivatives the students know at this point is very short—sums (with coefficients) of powers, exponentials, and natural logarithms. Now they will increase the list dramatically. You can help them to see this for themselves after introducing composite functions just by asking them each to list three examples of functions constructed by composition from the simpler ones.

The chain rule is motivated in the text by the problem of finding the cost of gas (dollars per minute), a rate, from two other rates, the price of gas (dollars per gallon) and gas usage (gallons per minute). Students usually know to multiply the rates, they say by common sense, which probably means that they can see that multiplying rates makes the units come out right. So there you have it, an example where one rate is the product of two others. Once you show them that the total cost of gas as a function of time is a composite function, they are ready for a full statement of the chain rule.

If you prefer to start with a numerical example, you might consider an example like $f(x) = x^2$, $g(x) = 3^x$, and $h(x) = g(f(x)) = 3^{x^2}$. Ask students to guess $h'(2)$, and they will generally try working with various combinations of $g'(x) = (\ln 3)3^x$ and $f'(x) = 2x$. Then compute $(h(2.001) - h(2))/0.001$ and compare with the prediction. It quickly becomes apparent that the derivative of a composite function is not as easy to guess as they might have thought.

But most of the class period should be spent with lots of examples and a good bit of drill. Help the students learn how to tell when they need to apply the chain rule, or what is the same thing, how to recognize a composite function. Give them some problems to work out at their desks to check that each one has really understood the technique. It is a good idea to have them evaluate some of the derivatives at specific points, and from time to time you might have them check such an answer numerically with a difference quotient (using their calculators). You don't want them to forget that every derivative they will ever see can be approximated by a difference quotient!

Once the chain rule is in place, it can be used to establish the formulas $\frac{d}{dx}(a^x) = (\ln a)a^x$ and $\frac{d}{dx}(\ln x) = \frac{1}{x}$ presented in Section 5.2. The proofs can be found in Section 5.6, but you may choose to mention them here instead.

Problems

A number of the routine problems (1-32) should be done in class, by the students, and a number should be assigned. The chain rule takes a while to become second nature. The application problems mainly involve exponential functions and are, as always, of utmost importance. Differentiation of periodic functions also makes use of the chain rule but is dealt with separately in Section 5.4.

Suggested Problems

#1, 5, 8, 11, 18, 25, 32, 34, 36, 38

5.4 THE PRODUCT RULE

One half to one class.

Key Points

Differentiation formulas for products of functions.

Ideas for the Class

Your first task may be to convince students that there is a need for a product rule, that the derivative of a product is not the product of the derivatives. Some will be convinced by consideration of $x^2 \cdot x^3 = x^5$ whose derivative equals $5x^4$ and not $2x \cdot 3x^2 = 6x^3$ which is wrong in both the coefficient and exponent. Some will prefer a graphical approach. For example, the graph of the surge function $x \cdot e^{-x}$ from Section 4.6 shows that it is increasing over part of its domain and decreasing over another part, so its derivative can not be $1 \cdot -e^{-x}$ which is negative everywhere.

A justification for the product rule is given in Section 5.6, which you can refer students to if they ask, or go over quickly in class. But you will want to spend most of your time with examples and more examples. Consider getting students up at the board, differentiating as you throw them functions. Or give them several one minute quizzes to do at their desk, with correct answers revealed to them immediately thereafter. This will help them know for themselves whether they really understand the product rule, and generate questions if they do not. Start simple, but work up gradually to products of composite functions. Your students are probably still just getting comfortable with the chain rule, and the more practice with it, the better.

For decades students have laughed at the description of the product rule as the HiHo rule, a la Hoagy Carmichael. "The derivative of HiHo equals Hi dee Ho + Ho dee Hi". They may enjoy this, and have fun identifying Hi and Ho in a couple of problems.

The quotient rule has not been included in the text. Students will need the derivatives of very few quotients, and they can be taught do them with the product and chain rules, since $u/v = u \cdot v^{-1}$. If you prefer, you can go ahead and teach the quotient rule.

Problems

Again a mixture of routine (1-18) and application problems (19-23) are advisable. Problems 19 & 20 will seem like old friends to most students. They should enjoy doing these. Problem 22 is an excellent problem on which sufficient time should be spent. Perhaps assign it to small groups and then bring everyone back together to talk about it - especially about the (b) part.

Suggested Problems

#1, 3, 4, 6, 10, 18, 19, 20

5.5 DERIVATIVES OF PERIODIC FUNCTIONS

Optional; if covered, one half to one class.

Key Points

Differentiation formulas for $\sin x$ and $\cos x$.

Ideas for the Class

Start class with the graph of a periodic function, say something with an interesting shape like $\sin x + \cos 2x$, or you might choose a zigzag piecewise linear periodic function. (Just sketch it on the board, don't bring formulas into the discussion.) Have the class discuss the function and sketch the graph of its derivative. The concept you want your students to discover is that the derivative of a periodic function is also periodic. You might assign them to explain why in a carefully written paragraph, with graphs.

Of course the most important periodic functions are sine and cosine. The formulas for derivative of sine and cosine are justified only very informally in the text. The students are led to see, for instance, that the graphs of $\cos x$ and of the derivative of $\sin x$ share qualitative features, and a difference quotient approximation for the derivative of $\sin x$ at $x = 0$ is compared with the value $\cos 0 = 1$. This should be enough for most students.

If you have a strong class, you could have them graph the function

$$\frac{\sin(x + h) - \sin x}{h}$$

on a graphing calculator for several values of h such as 1, 0.5, 0.1, and 0.01. This reminds students that derivatives always go back to difference quotients, and comparison with the graph of cosine makes a very strong case for the derivative formula.

A good illustration of the chain rule is gotten by comparing the graphs of $\sin x$ and $\sin 3x$. Students will see that the graph of $\sin 3x$ gets steeper, and so they can predict that its derivative gets higher. This matches what comes from the derivative formulas.

Problems

Having been familiarized with the chain rule in Section 5.3, students should feel comfortable with differentiating periodic functions. Again a mixture of routine problems (1-16) and application problems (18-20) are advisable. Emphasize Problem 18, especially the interpretation of the function and derivative values. Discuss the (b) part of Problem 20. The intuitive idea here is expanded in Chapter 6.

Suggested Problems

#1, 2, 7, 12, 17, 18, 19

5.6 VERIFYING THE DERIVATIVE FORMULAS

One class.

Key Points

Proving derivative formulas with difference quotients.

Ideas for the Class

It is time for the students to see how some of the derivative formulas that they have been using can be proved. This is your chance also to remind them that, in the end, everything there is to know about derivatives goes back to the definition, which is in terms of difference quotients. Keep it simple. Students in this course can usually follow simple algebraic arguments, but it is often best to work up to the algebra through some preliminary concrete computations. You have to convince them first that algebra is their friend.

For example, suppose you want to prove that the derivative of x^2 equals $2x$. You might begin with a classroom project of making a huge table of approximate values by computing tons of difference quotients. Do, or have students do, $\frac{(2+0.1)^2 - 2^2}{0.1}$, then $\frac{(3+0.1)^2 - 3^2}{0.1}$, then $\frac{(4+0.1)^2 - 4^2}{0.1}$, etc., approximations for the derivative at 2, 3, and 4. Students will begin to find this tedious and repetitive, and so they will be glad to hear that they can do an infinite number of these computations all at once — it's the power of algebra!

So go back to redo and extend your table but using the formula $\frac{(x+0.1)^2 - x^2}{0.1} = 2x + 0.1$. But the table is only approximate, and we want to improve its accuracy. Start over again, this time with $\frac{(x+0.01)^2 - x^2}{0.01} = 2x + 0.01$, then again with $\frac{(x+0.001)^2 - x^2}{0.001} = 2x + 0.001$.

Your students will be begging for the chance to do all values of x and all values of h at once with more algebra: $\frac{(x+h)^2 - x^2}{h} = 2x + h$. And that is really all there is to it. They can all now see that the smaller h is, the closer the difference quotient is to $2x$, and so they have their first proof.

A proof of the product rule is included in this section, but you may choose to omit it.

Problems

Problems 1-7 are simple applications of the definition of a derivative. Problems 8 & 9 test whether the student has a graphical understanding of this definition. One of these can be done as a group activity in class and the other as an assigned problem. Students might need guidance on Problem 10. Sketching the graph of $f(h)$ might seem like a strange activity.

Suggested Problems

#3, 5, 9

5.7 FINDING ANTIDERIVATIVES

One class.

Key Points

Antiderivatives. The indefinite integral. Use of antiderivatives and the Fundamental Theorem to evaluate definite interals.

Ideas for the Class

Students should be introduced to antiderivatives and their notation as indefinite integrals, since they may come across them in another course. With some simple antiderivatives available they can take another look at the Fundamental Theorem, seeing how it is used to evaluate a few definite integrals exactly, namely those for which there are formulas for both the integrand and the antiderivative. But keep it very simple. Don't get carried away with fancy techniques of antidifferentiation. In the applications in the rest of the text, all integrals are definite integrals and so can be evaluated if need be by Riemann sums or by estimating areas on graphs.

Problems

Assign a number of the three sets of routine problems (1-10, 11-18, 19-26). Although these problems are all simple as far as integration goes, do not overestimate the students' ability on finding antiderivatives. Antiderivatives are a notoriously weak spot in the make-up of the average student. Problem 30 is a good problem for developing overall understanding.

Suggested Problems

#1, 6, 8, 11, 15, 16, 17, 19, 23, 24, 30

CHAPTER SIX

Five to seven classes.

Overview

In this chapter students will see how the derivative and definite integral can be used in a variety of settings. The analytical tools developed in Chapter 5 are brought into play, but the emphasis throughout remains on the concepts themselves and their graphical and verbal interpretations.

The derivative is used to explore the shapes of graphs given by formulas. Optimization is touched on briefly. The relationship between maximum profit and marginal cost and revenue is developed more fully than was possible in Chapter 2. Average cost is studied graphically and a connection is made with marginal cost. The primary use of the definite integral is to compute total change from rate of change. Other applications of the definite integral discussed are to bioavailability of a drug and to consumer and producer surplus of demand.

No student should have to leave this course wondering whether there are any applications of calculus to the real world.

6.1 CRITICAL POINTS AND INFLECTION POINTS

One ot two classes.

Key Points

The derivative: increasing/decreasing functions, critical points, and local extrema. The second derivative: concavity and inflection points.

Ideas for the Class

This section is about using calculus to uncover the relationship between equations and graphs.

Students may think at first that with function-graphing technology they don't need these ideas. Certainly they will not see the point of laboriously using calculus to graph fifteen functions by hand that can be graphed easily on a machine. But the first example in the text illustrates one way we can use calculus to lessen our frustration with the technology, namely to help us find a suitable calculator window for an unfamiliar graph. A couple of other examples in the same vein to try is: $y = -7x^3 + 23x^2 - 19x - 12$ or $y = xe^{-3x} + x^2 + 50$. (You might remind students that they can look for zeros of the derivative with their calculator, too!)

Another approach to this material that admits from the start that graphing technology is available is to start with *both* a formula and its graph. Ask the students first to describe what they see in the graph — intercepts, increasing/decreasing sections, concave up/down sections, critical points, inflection points, local maxima and minima, asymptoptes. Then ask where all these visual features of the graph are buried in the formula. They should be able to uncover them by analyzing the function and its first and second derivatives. If you discussed the surge function in Section 4.6, you will definitely want to discuss Example 5 on critical points and inflection points of $y = xe^{-x}$.

Make sure students are confidently thinking geometrically to distinguish a local minimum from a local maximum. If a derivative changes from negative to positive then the function graph must first go down then back up, so the transition point must be a local minimum. To ensure that your students really see this picture rather than trying to memorize a 'test' for local minima, you might ask them to write a paragraph with illustrations. Show them the graph of a derivative crossing the x-axis and have them sketch a tiny piece of the function at x-values near the crossing.

Students should already be comfortable with the idea that the sign of the derivative is related to the monotonicity of the function, but they may not know how to use it in a purely mathematical setting, especially where the function is defined by a formula. For example, you might tell the students that one of the functions $f(x) = x^3 + 150x^2 + 8000x$ and $g(x) = x^3 + 150x^2 + 7000x$ is always increasing and the other is not and ask them to tell which is which. The derivatives tell you: $f'(x) = 3x^2 + 300x + 8000$ has no real zeros while $g'(x) = 3x^2 + 300x + 7000$ has two, so the first cubic is always increasing.

Keep the focus as much as possible on graphs.

Problems

In Problem 5 the notation might be a problem. Discuss with the students what is given in this problem and they might not experience any difficulty at all. Similarly, when dealing with Problems 6 & 7, make sure that students realize that they are looking at f' and not f. Of the technical problems (8-13) at least one should be assigned and discussed. Problems for interpretation are 14, 15, 24, 26, 33, 34, 39 & 40. Numerical problems are 21 & 22. Graphical problems are 1-4, 6, 7, 20, 23, 27, 28-31 & 35-38. Problems requiring more algebraic expertise are 17-19 and 25. Students would want to do problems from each of these groups of problems.

Suggested Problems

#1, 5, 6, 9, 19, 34, 39, 40

6.2   OPTIMIZATION

One class.

Key Points

Global extrema. Maximizing profit. Marginal cost and revenue.

Ideas for the Class

Students already know that local maxima are found at critical points, and so they will easily understand that to find a global maximum they need to just search through the list of critical points (and endpoints of domain if any) for the one that gives the greatest value. You can either just make this point with one or two examples, or drill a bit, depending upon how important working with formulas is for your class.

Of more interest to some students will be the discussion of profit maximization that is a follow-up to Section 2.6, Marginal cost and Revenue. Example 3 is worth discussing in detail. In particular, note that in the graph of Figure 6.34 there are *two* production quantities at which marginal cost equals marginal revenue, a situation discussed often in economics classes. Students should be able to explain clearly which of the two points corresponds to maximal profit and why. You might try graphing part of a similar graph, showing only one crossing of marginal cost and revenue, and asking the students to make a recommendation of whether to increase or decrease production from the crossing point.

When discussing maximization of revenue, make sure that the students understand that it is maximization of profit that is really of interest. Examples have been chosen for the text (bus company, amusement park) where the cost is thought of as independent of the quantity of sales, and so in these cases maximum profit and maximum revenue occur at the same point.

Problems

The first four problems practice skills. Do not underestimate the importance of this. Problems 6-15 involve cost and revenue. Explanations should be emphasized. Problems requiring graphical interpretations such as 9 & 10 are suited for generating discussions. Problem 11 is interesting in so much that the cost function is explicitly given as a cubic polynomial. Encourage students to graph this function as well as the revenue function. This problem would be good for group work. Problem 16 is based on the Cobb-Douglas function which is discussed extensively in Chapter 7. Assign this problem only to students who would like enrichment work.

Suggested Problems

#1, 6, 8, 10, 13

6.3   MORE OPTIMIZATION: AVERAGE COST

One class.

Key Points

Average cost and its graphical interpretation. Minimal average cost occurs when marginal cost equals average cost.

Ideas for the Class

This section is really an extended example. It demonstrates the power of calculus and graphs to reveal hidden relationships between two different concepts. In this case, it is a relationship between average cost and marginal cost. The link is made graphically via slope. Students will see graphically that average cost of production is the slope of a line from the origin to the cost curve, and they already know that marginal cost is the slope of a tangent line to the cost curve. You might ask them if there is a production level where the two lines are the same. If they can answer the question and explain its significance, they will have understood the point of this section of the text.

Problems

A number of problems are based on graphs (1, 3, 4, 5, 8 & 11). The aim here is mainly for the student to visualize the difference between marginal cost and average cost. Most of the other problems (6, 7 & 10) require the formulas for marginal cost and average cost. Students often find this much simpler than visualization.

Students who did the Cobb-Douglas problem in the previous section would like to do Problem 9 as well. Problem 12 is a good group work problem. It might be good advice to first try out this rule with actual functions such as in Problem 6.

Suggested Problems

#1, 3, 4, 5, 7

6.4   APPLICATIONS OF THE DEFINITE INTEGRAL

One class or two classes.

Key Points

Using the Fundamental Theorem to sketch the graph of f given the graph of f'. Integration of rates in a variety of applied contexts.

Ideas for the Class

Students have been working in preceding sections mostly with derivatives, so it is a good idea to remind them of the basic interpretations of the definite integral: average value of a function; area under a graph; and most important, total change of a quantity computed from its rate of change. The primary tools students have for computing definite integrals are all approximate: estimating areas on graphs, Riemann sums, evaluation with computer or calculator software. (Exact evaluation using the Fundamental Theorem and antiderivative is not emphasized in this text, though it has been discussed in Section 5.7 and can be used if desired.)

Class time is best spent with lots of examples. If you want your students to work a large number of the problems, a second day of discussion would be appropriate. Note that Examples 3, 4, 5, and 6 emphasize the total change from rate of change (Fundamental Theorem) interpretation of the definite integral. Another rate of change problem that generates an excellent class discussion is the following:

Example 1   Figure 6.4.1 is the graph[1] of the rate r (in arrivals per hour) at which patrons arrive at the theater in order to get rush seats for the evening performance. The first people arrive at 8 am and the ticket windows open at 9 am. Suppose that once the windows open, people can be served at an (average) rate of 200 per hour.

Use the graph to find or provide an estimate of:

(a)   The length of the line at 9 am when the windows open.
(b)   The length of the line at 10 am.
(c)   The length of the line at 11 am.
(d)   The rate at which the line is growing in length at 10 am.
(e)   The time at which the length of the line is maximum.
(f)   The length of time a person who arrives at 9 am has to stand in line.
(g)   The time at which the line disappears.
(h)   Suppose you were given a formula for r in terms of t. Explain how you would answer the above questions.

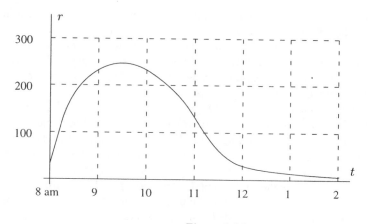

Figure 6.4.1

---

[1]From Calculus: The Analysis of Functions, by Peter D. Taylor (Toronto: Wall & Emerson, Inc., 1992). Reprinted with permission of the publisher.

The final application in the section, bioavailability of a drug, is a good one in part because to gain full understanding it is necessary to go back to Riemann sums. Students will see that the individual terms in the Riemann sum measure bioavailability for short time intervals, and that the integral computes the total for all the short subintervals. For another example requiring the same type of analysis, ask students to compute the number of man-hours required to complete a year long project, given a graph of the (fluctuating) number of workers on the job as a function of time over the course of the year. They should be able to explain using Riemann sums why the area under the graph gives the answer.

Problems

The first four problems require graphical interpretations. As always a number of these should be assigned to improve visualization. This is followed by a number of problems which require estimations or calculations (5-12 & 14). Where possible a graphical interpretation is required as well. Problem 13 is the only problem based on a table and so it is good for refreshing this aspect. Problems 15-21 are all of a biological nature and mostly require graphical interpretations. Students like these problems and also like discussing them. These make good group work and discussion problems.

Suggested Problems

#1, 3, 5, 7, 9, 13, 15, 17, 19

6.5 CONSUMER AND PRODUCER SURPLUS

One class.

Key Points

Supply and demand curves, consumer surplus, producer surplus, and the effects of wage and price controls.

Ideas for the Class

You might open class with a discussion of the nature of trade. When two persons trade items, what are the relative values of the items? Often their "common sense" tells students that two persons trade items of equal value. But that is not correct! Students must understand that when one person buys from another, each person winds up with more than they had before and so both persons gain from the trade (which is why they are willing to trade). This may seem paradoxical to students, but a simple example can help make the point.

Suppose I plan to pay up to $20 for a compact disc that I find in a store for sale at $15. The disc is worth $20 to me but I have traded for it with only $15, so I have gained $5 from the trade. Suppose the store would have been willing to sell for anything above $12. Then the disc is only worth $12 to the store, but they traded it for $15, so the store has gained $3 from the trade. And this is how trade enriches a country.

Once you are sure the students understand how people gain from single trades, you can proceed to the goal of this section, which is finding the total gains from trade (the consumer surplus) of a group of consumers who place different values on the same good or the total gains from trade (the producer surplus) of a group of suppliers who place different values on the same good. Do your students see that the demand curve keeps track of the way consumers value the good? Try asking them what the demand curve would look like if all consumers valued a good at less than $20. (We want them to say that the demand curve must intercept the price axis below $20.) Or how would the demand curve change if an advertising campaign caused consumers to value the good more highly? (The same quantities could be sold at higher prices, so the new demand curve would be higher than the old one.)

One final caution is in order here. It is easy for students to memorize the final formulas for consumer and producer surplus. But it is important (and much more interesting) to emphasize the modeling and thought processes that go into discovering the formulas. The applications to wage and price controls in the text and exercises are interesting and can help you to make the point that calculus can be a thinking tool just as much as a computational tool.

Problems

Most students find this section somewhat difficult. The section itself requires careful reading and the problems, although not many, will take time to do. Problems 1 & 2 are excellent for checking whether the students have grasped the concepts. Spend time on this. Problems 4 and 7 are good group work problems.

Suggested Problems

#1, 2, 4, 5, 6

CHAPTER SEVEN

Six classes.

Overview

Quantities of serious interest in the real world usually depend upon more than one variable. To model them mathematically we need functions of two or more variables. This final chapter is a brief introduction to such functions and their use.

The Rule of Four still applies. Two variable functions can be described numerically, graphically, algebraically, and verbally. Numerical representation is by rectangular tables of values. Graphical representation is by contour diagrams. After discussing these representations, partial derivatives are defined and interpreted as rate of change in a variety of contexts. The chapter closes with a brief introduction to constrained optimization and the method of Lagrange multipliers.

7.1 UNDERSTANDING FUNCTIONS OF MANY VARIABLES

One class.

Key Points

The Rule of Four: representing functions of two or more variables by contour diagrams, tables of values, formulas, and words.

Ideas for the Class

Prior student experience with functions of two or more variables is likely to be almost entirely verbal, so it is good idea to begin class in the verbal mode. The opening theme is that many-variable functions are all around us. Even functions previously studied as one-variable functions can be taken as examples. For instance, the balance in a bank account earning interest continuously depends on the time it has been earning according to the formula $B = Pe^{rt}$, where we regarded initial investment P and interest rate r as constants to make this a function of the single variable t. In actuality, r and P are also variables, so we have a function of three variables $B(P, r, t) = Pe^{rt}$.

Some more examples of functions of several variables:

- The chirping activity C of crickets depends on the time of day t and the temperature T, so that $C = f(t, T)$.

- The temperature T can be expressed in terms of location with latitude a and longitude b, so $T = h(a, b)$.

- Your income tax due S depends on your income I, your exemptions e, and your deductions D, so $S = g(I, e, D)$.

You can get many other examples from the students. After they suggest a few, you might ask them if they can think of a quantity that depends on only a single variable, and they will not find it easy to do so.

In reality, most quantities depend on many variables. For example, the number N of years it takes to complete a college degree depends on (among other things) the major M, the choice of school s, monetary support m, academic background A, the time taken to choose a major t, intelligence I, and family support F, so $N = f(M, s, m, A, t, I, F)$. The more variables that are involved, the more difficult it is to understand the function's behaviour completely.

Functions of two variables are represented graphically in this text by contour diagrams, not by surface graphs. The first example, the weather map in Figure 7.1, will be familiar to students, though they will not have associated the word 'function' with it. Contour diagrams will be discussed in detail in Section 7.2, so you may want to spend most of your time in this section with general discussion and tables of values.

You might put up a transparency of Table 7.1 (beef consumption) and ask, "What observations can you make?" Such an open-ended question should allow for discussion of such issues as: as income I increases, consumption C increases; as price p increases, C decreases. If you ask, "How can you tell?" students automatically explain by looking at one column or row of data at a time, leading naturally to the notion of cross-section (holding one variable constant and letting the other vary). Let them know that they have discovered for themselves one of the ways in which one-variable mathematics can be used to study more complicated functions.

You might select one of the functions for money M in a bank account as a function of initial balance and time given in Example 2 and have groups of students compute values, eventually putting together a table of values as a class. Again ask, "What observations can you make?" They should see this time that as either variable increases, M increases. Ask whether they could have predicted it in advance.

Now ask whether the function $f(x,y) = x - 2y$ is increasing or decreasing. A discussion will ensue, and students will begin to see that the language of one variable functions will sometimes have to be refined for it to make any sense at all in a two-variable setting. In this case, they might say that f is increasing with respect to x and decreasing with respect to y.

Don't take tables of values completely for granted. If you ask for a table of values for $f(x,y) = x - 2y$, you are likely to find a few of the students using the two column format they are familiar with — in the first column a list of points (x,y) and in the second column the values of the function. Rectangular tables are so natural to us that we tend to forget that it is a step requiring some real insight to appreciate the advantages of arranging the table of values in rectangular form. This can be brought home to the students if you wish by handing them a two-column table of values of a two variable function and asking them to use it to answer some simple questions, even as simple as evaluating the function or determining how the value changes when one of the independent variables increases. They will all express a preference for the rectangular format. You might ask your students to prepare a table of values both ways and comment on the differences.

Problems

Problems 1-3 as well as Problems 4-9 refer back to examples in the section and could be done in little groups in class. These problems are all intended to make the student comfortable with the notion of a function of many variables. Interpretations and visualizations are extremely important. Problems 10-12 are on increasing and decreasing functions in one variable. Problems 13-21 are all relatively simple but aim to increase understanding. Problems 22-26 form a unit and could be given as a group project. Similarly, Problems 27-30 form a unit and are suited as a group project.

Suggested Problems

#11, 12, 16, 22-26

7.2 CONTOUR DIAGRAMS

One class.

Key Points

Reading and interpreting contour diagrams for functions of two variables. Contour diagrams and tables of values. Finding contours algebraically.

Ideas for the Class

What are some of the desirable properties of a good pictorial representation of a function of two variables? You might pose this question to your class as a discussion topic, possibly with reference to a weather map such as Figure 7.1 which is the example of such a pictorial representation that they are most familiar with. If your students seem slow in responding, you can get them started by asking them to name some advantages of graphs over formulas for one variable functions. Some of the properties of a good picture of a function that they could mention are as follows:

- It should be possible and even easy to evaluate the function from the picture.
- The picture should give at a glance a good global idea of where the function has high values and where low values.
- It should be easy to determine from the picture how function values change (either increasing or decreasing) when one of the variables is allowed to vary.

In this text, the pictorial representation chosen is the contour diagram. Contour diagrams are widely used in the applied literature precisely because they meet the criteria mentioned above. The task of Section 7.2 is: (1) to get students comfortable reading and interpreting contour diagrams by looking at examples from a variety of contexts; and (2) to explore the interplay between different ways of representing the same function, tables of values, formulas, contour diagrams, and words. Spend some time on the significance of the spacing of the level curves. What can you conclude if the level curves are very close together? or far apart? or equally spaced? or unequally spaced? Note that it is not necessary or even always helpful to interpret contour diagrams as giving level curves of a surface.

It is not easy to draw a contour diagram given a formula for a two variable function. This is not yet done or not done well by calculators and small computers. For all but the simplest examples you will want to come to class with previously prepared diagrams to display or distribute.

A different way to present a two-variable function pictorially, very common in the scientific literature, is to graph several sections of the function on the same coordinate plane. An example is given as Problem 21 of Section 7.1.

Students may ask whether the contour diagram is the graph of the function. The answer of course is no. The graph is a surface in three dimensional space that mathematicians, with their geometric insight, are very attached to. However, the graph is not so useful as a primary way to present a function because it is not possible to read function values from a two-dimensional rendering of the surface. The graph of a two-variable function is not introduced in the text.

Problems

These problems relate contour maps to surfaces via graphs, formulas and tables. Hardly any calculations are required. The emphasis is on explanations and interpretations. Students might need a bit of guidance on starting Problems 9-14 but they should soon see the simplicity of it. Do not assign too many problems. The concept of contour diagrams is probably different to any other mathematics that the students have done so far, so give them time to digest it.

Suggested Problems

#2, 3, 8, 9, 20, 23

7.3 THE PARTIAL DERIVATIVE

One class.

Key Points

Definition of partial derivatives; approximating them with difference quotients given a contour diagram or table of values; using units to interpret partial derivatives in applied contexts.

Ideas for the Class

Open class by asking for the rate of change of a particular two variable function at a point with respect to changes in one of the variables. You can do this with either a table of values or a contour diagram (for example the heat-index table in Table 7.2 or contour diagram in Figure 7.21 on page 387) or even with a formula. With possibly a little reminder of the concepts from Chapter 2 , your students will discover the partial derivative for themselves. They are already comfortable varying one variable at a time to see whether it causes the function to increase or decrease. They just have to quantify this process by combining it with a difference quotient.

The rest of the class should focus on the practical meaning of the partial derivative in applied contexts. The difference quotients show what the units of the partial derivative are, and it is interesting that the two partial derivatives can have completely different units and hence completely different meaning.

Finally, it never hurts to remind your students that a difference quotient is not the partial derivative; it is an approximation for the partial derivative. This, however, does not diminish the usefulness of the difference quotient.

Problems

The concept of the partial derivative will be new to the students, so they will need to spend time on the interpretation. These problems require hardly any calculations, only interpretations. As previously, be picky as to how they formulate answers. Do a number of these problems in class. Get them to verbalize but also to write down interpretations. Although they might intuitively feel that they understand what a partial derivative is, the skill of verbalizing should be developed. Problems are approached from the graphical side (7, 9, 21 & 22), the numerical side (6, 14-16) and the rest from a broader interpretation aspect. Problem 25 is suitable as a group project or as an enrichment problem.

Suggested Problems

#1, 6, 7, 9, 12

7.4   COMPUTING PARTIAL DERIVATIVES ALGEBRAICALLY

One class.

Key Points

Algebraic computation of partial derivatives. The Cobb-Douglas production model.

Ideas for the Class

Students will have no difficulty computing partial derivatives algebraically. You just have to make sure that they connect what they are doing with the concepts of the preceding section. This can be done by asking often for the values of partial derivatives at a point, so they will have to evaluate the partial derivative function. On the one hand, they can compare the value they get this way with a difference quotient approximation. You could even encourage them to use difference quotient approximations this way as a check on their algebraic computations. On the other hand, in practical situations students can be asked to interpret the value of a partial derivative in the context of the problem, as in Example 2.

The last part of the section presents the Cobb-Douglas production model for productivity as a function of the two variables capital and labor. It has both historical and practical interest.

Problems

Students experience no difficulty with the technique of finding partial derivatives. Problems 1-14 are practice exercises from which a number have to be done. Problems 15-20 combine the technical skills with interpretations and is of course what it is all about. Problems 21-25 all concern Cobb-Douglas functions. Problem 23 and one or two of the others could be valuable.

Suggested Problems

#1, 4, 15, 16, 18, 23, 24

7.5   CRITICAL POINTS AND OPTIMIZATION

One class or less.

Key Points

Local maxima and minima of functions of two variables. Critical points.

Ideas for the Class

Students used to reading topographical maps will know that local maxima and minima show up on contour diagrams surrounded by closed contours. That these local extrema can only occur where both partial derivatives equal zero is an easy analog of the one variable criterion for local extrema.

Finding critical points algebraically from a formula for a two-variable function is not necessarily easy or even possible. It requires solving a system of (often nonlinear) equations. You may not want to ask your students to do much of this by hand.

Saddle points are not discussed in the text, but they are critical points. Ask your students for a contour diagram representing a mountain pass. You may want to show them a simple example such as the function $f(x,y) = x^2 - y^2$ at the origin.

Problems

The first two problems are good problems to do in class. Problem 2 could generate some discussion. Problems 3-6 are necessary practice problems. Problems 7 & 8 are graphical problems which the students will enjoy doing. Assign at least one of them. Problem 10 requires careful thinking and is suitable for group work.

Suggested Problems

#2, 3, 8, 10

7.6   LAGRANGE MULTIPLIERS

One class.

Key Points

Extrema of functions of two variables subject to a constraint. Method of Lagrange multipliers. The meaning of the multiplier.

Ideas for the Class

Constrained optimization is a multivariable concept. It has no analogue in one variable mathematics, so it is wise to spend some time on the difference between unconstrained and constrained optimization of a two variable function. You might show a huge table of values or a contour diagram and ask for the point corresponding to the greatest value of the function. Then introduce a constraint by drawing a curve right across the table or diagram and ask for the point on the curve corresponding to the greatest value of the function. Find the highest point anywhere in North America, or constrain to the United States-Canada border. Place a mound of playdough on a table and find the highest point, then make a vertical cut with dental floss and find the highest point on the cut. Consider Example 1, a linear budget constraint, for the analytic view.

An in-class discovery activity suitable for working in groups goes as follows. Distribute contour diagrams for a number of functions. Students draw some constraint curves of their own choosing on the diagrams, and search for geometric properties of the maximal points on the constraints. They will find that the maximal points can be only where the constraint curve is tangent to a level curve of the diagram. Now go back to the playdough, and draw the level curve round the the playdough that goes through the maximal point on the cut. Tangency again, this time of the cut and the level curve.

Students know that tangents are related to derivatives, and so will be ready for the statement and use of the method of Lagrange multipliers. There is no formal derivation of the method in the text.

Note that the Lagrange multiplier λ has a practical interpretation, which is explained in the text. It tells how the maximal value changes if the constraint is changed slightly. A little more formally: if m is the maximal value of $f(x, y)$ subject to the constraint $g(x, y) = c$ and M is the maximal value of $f(x, y)$ subject to the constraint $g(x, y) = c + \Delta c$, then $M \approx m + \lambda \Delta c$ (if Δc is small enough). Use of this interpretation of λ is illustrated in Example 4.

Problems

Students firstly have to familiarize themselves with the terminology and the technique of Lagrange Multipliers. Practice problems such as 1-5 will be helpful. By now students should also be familiar with Cobb-Douglas functions and ready to work on applied problems. A graphical problem such as Problem 8 is valuable and the numerical problem (16) is also advisable. One or two other problems added to this will provide enough exercise in applying the technique of Lagrange Mulitipliers.

Suggested Problems

#1, 6, 8, 13, 16

PART III

SAMPLE SYLLABI

Sample Syllabus (University of Arizona)

Lect.	Text/Section	Lect.	Text/Section
1	Intro, 1.1	23	4.3
2	1.2	24	4.5 and/or 4.6
3	1.3	25	REVIEW
4	1.4	26	TEST II
5	1.5	27	5.1
6	1.6 (1.7 — optional)	28	5.2
7	2.1	29	5.3
8	2.2	30	5.4
9	2.3	31	5.5
10	2.4	32	5.6
11	2.5	33	5.7
12	2.6	34	6.1
13	REVIEW	35	6.2
14	TEST I	36	6.3
15	3.1	37	6.4
16	3.2	38	REVIEW
17	3.3	39	TEST III
18	3.4	40	6.5
19	3.5	41	REVIEW
20	3.6	42	REVIEW
21	4.1	43	EXAM
22	4.2		

Sample Syllabus (University of Pretoria)

Schedule: Spring 1997

Week	M	Tu	W	Th	F	Sa
Jan 27-31	—	Intro	F	—	F	—
Feb 3-7	F	F	F	—	F	—
Feb 10-14	F	C §1.1	C §1.2	—	C §1.3	—
Feb 17-21	F	C §1.4	C §1.5	—	C §1.6	—
Feb 24-28	F	C §1.7	C Rev. Ch. 1	—	C §2.1	—
Mar 3-7	F	C §2.2	C §2.3	—	C §2.4	—
Mar 10-14	F	C §2.5	C §2.6	—	C Rev. Ch. 2	Exam 3/15
Mar 17-21	F	C §4.1	C §4.2	—	Pub. hol.	—
Mar 24-28	F	C §4.3	C §4.4	—	Pub. hol.	—
Mar 31 - Apr 4	Pub. hol.	—	—	—	—	—
Apr 7-11	—	—	—	—	—	—
Apr 14-18	—	C §4.5	C §4.6	—	Rev. Ch. 4	—
Apr 21-25	F	C §5.1	C §5.2	—	C §5.3	—
Apr 28 - May 2	Pub. hol.	C §5.4	C §5.5	Pub. hol.	—	—
May 5-9	F	C §5.6	Prob. sheet	—	Prob. sheet	—
May 12-16	F	C §3.1	C §3.2	—	C §3.6	Exam 5/17
May 19-23	C §5.7	C §3.3	C §3.4	—	C §3.5	—
May 26-30	Rev. Ch. 3	Rev.	Rev.	—	—	—

Remarks:

1. The course consists of Calculus (C) and Finite Mathematics (f). We started off with Finite Mathematics while we were waiting for the newly published textbok *Brief Calculus* to arrive.

2. The first five chapters of this textbook were covered in the course. The last two chapterrs will form part of the follow-up course in the second semester.

3. The strange sequence in which Chapter 3 was taught was due to the fact that students did not have access to technology for calculating definite integrals. After doing Riemann sums, we did the Fundamental Theorem and antiderivatives before proceeding with areas, average values and interpretation of the definite integral (3.3-3.5). Strange as it may sound, this was very successful and can be recommended.

PART IV

MASTERS FOR OVERHEAD TRANSPARENCIES

Transparency Master for Section 1.6

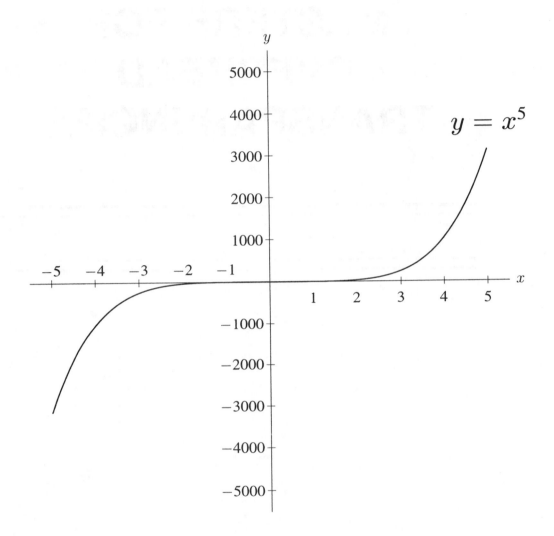

Transparency Master for Section 1.6

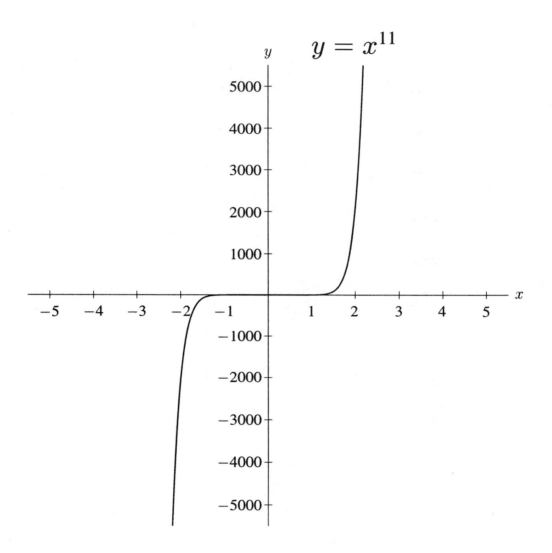

$$y = x^{11}$$

Transparency Master for Section 1.6

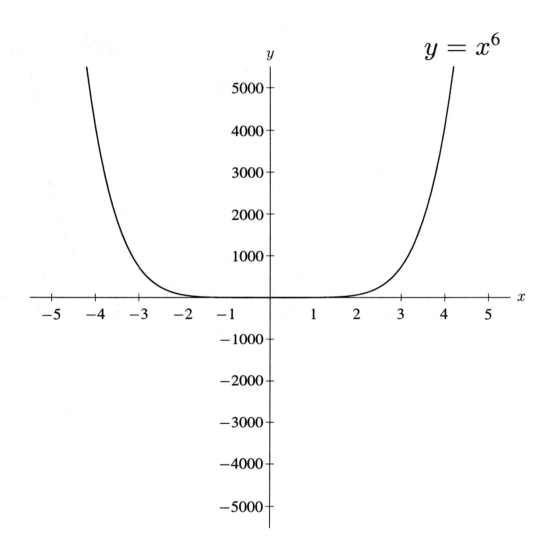

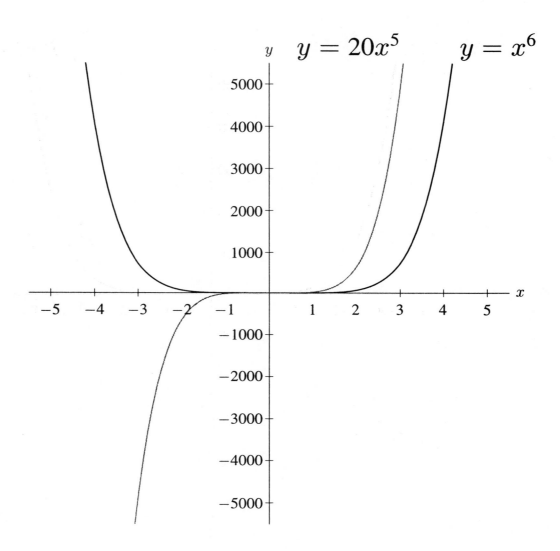

Transparency Master for Section 1.6

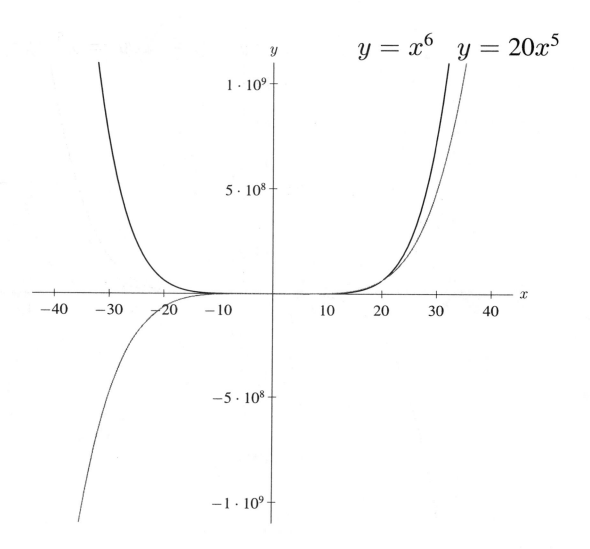

Transparency Master for Section 4.1

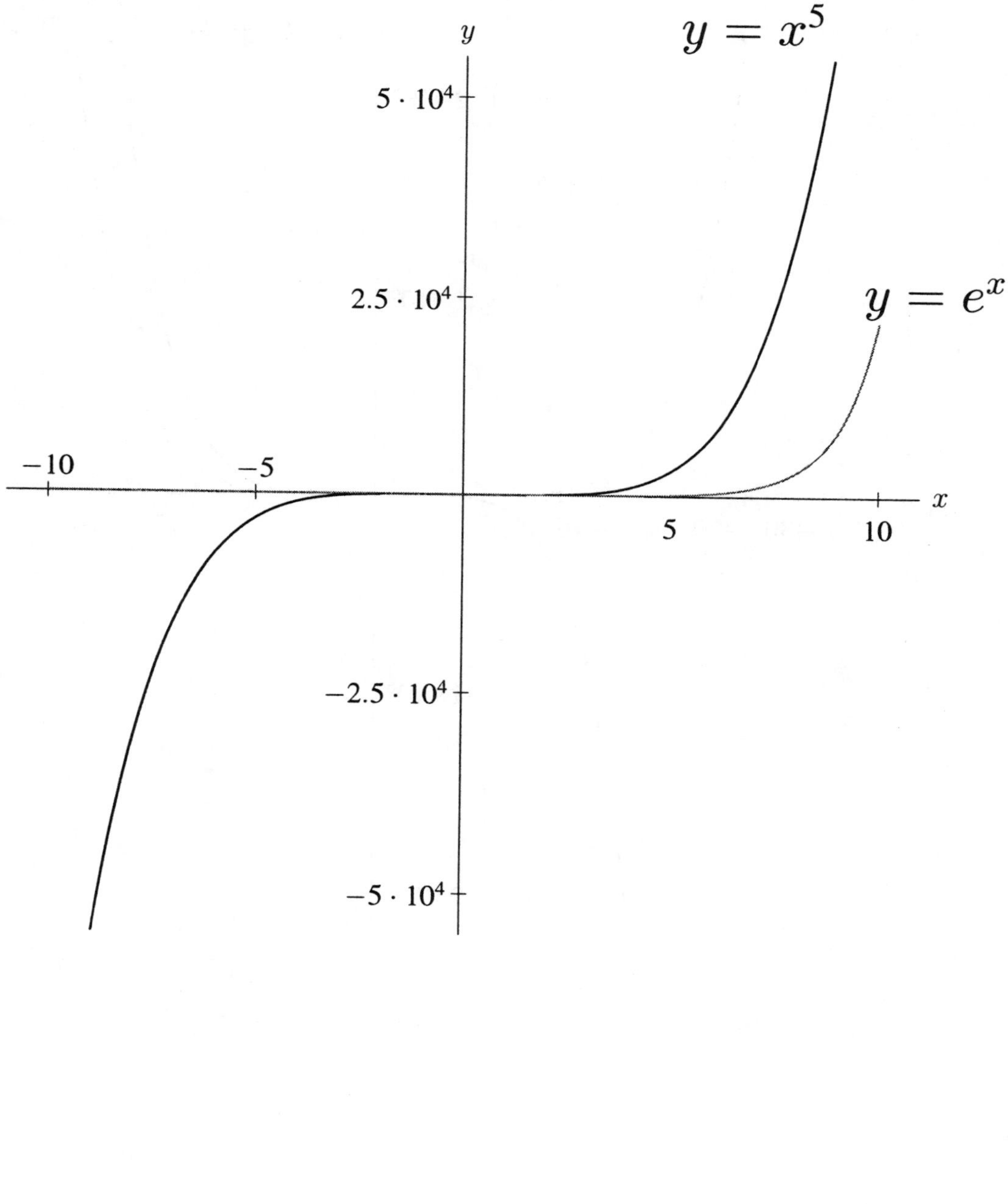

Transparency Master for Section 4.1

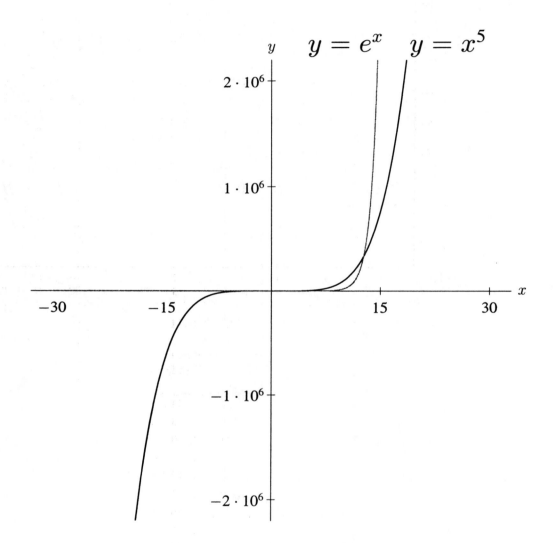

Transparency Master for Section 4.4

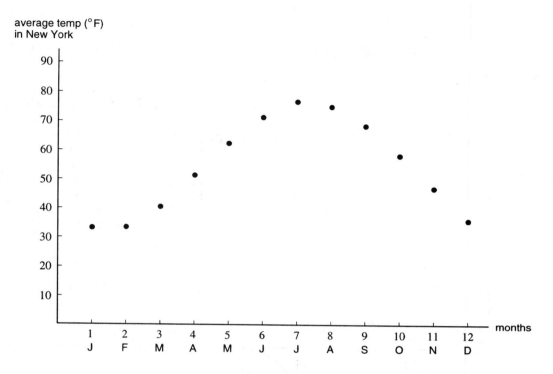

average temp (°F) in New York

TABLE 0.0.1

Average Temperature (°F) in New York by Month

Month	
Jan	33.2
Feb	33.4
Mar	40.5
April	51.4
May	62.4
Jun	71.4
July	76.8
Aug	75.1
Sept	68.5
Oct	58.3
Nov	47.0
Dec	35.9

Transparency Master for Section 2.3

The graph of f is given. Sketch the graph of f' on the axes below.

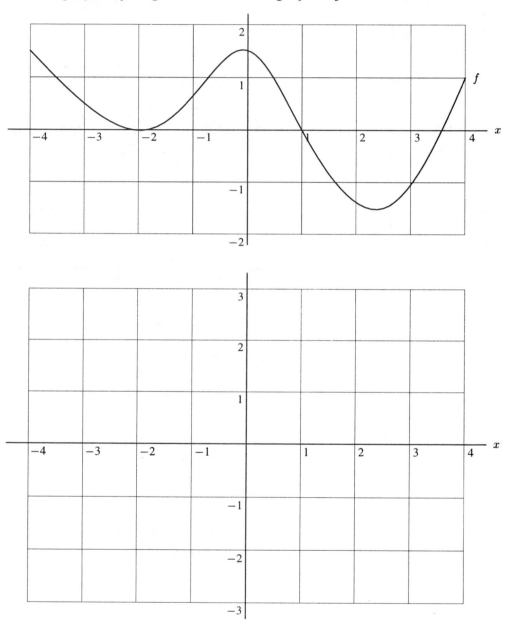

TABLE 0.0.2 *Quantity of beef bought (pounds/household/week)*

$I\backslash p$	3.00	3.50	4.00	4.50
20	2.65	2.59	2.51	2.43
40	4.14	4.05	3.94	3.88
60	5.11	5.00	4.97	4.84
80	5.35	5.29	5.19	5.07
100	5.79	5.77	5.60	5.53

Beef consumption in the US
(in pounds per household per week)

Transparency Master for Figure 7.16 in Section 7.2

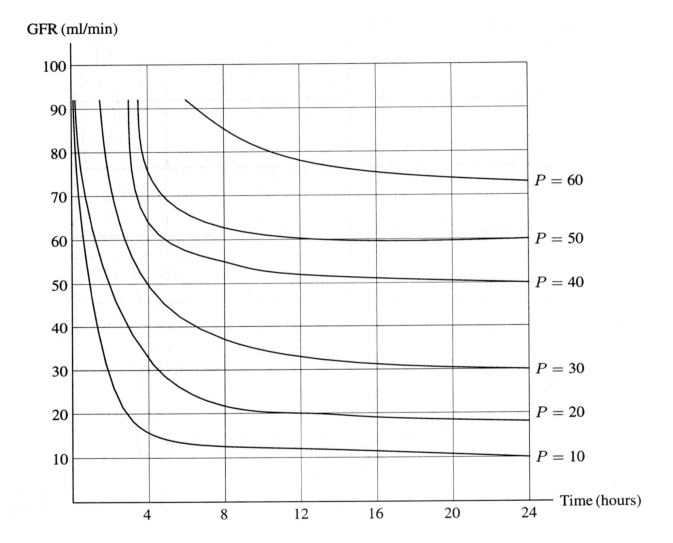

Transparency Master for Figure 7.24 for Problem 1, Section 7.2

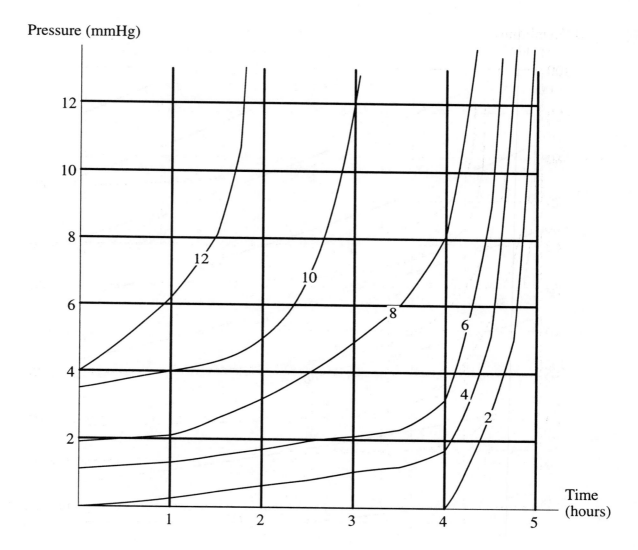

Transparency Master for Figure 7.30 for Problem 8, Section 7.2

loan
amount($)

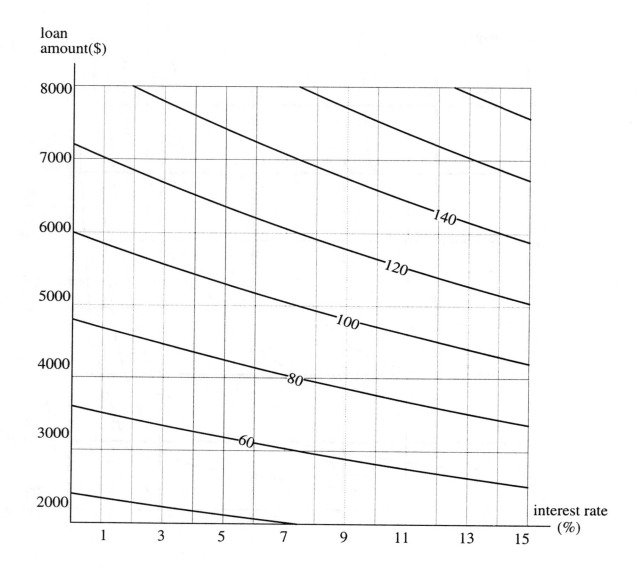

Transparency Master for Figure 7.39 for Problem 26, Section 7.2

kilometers north

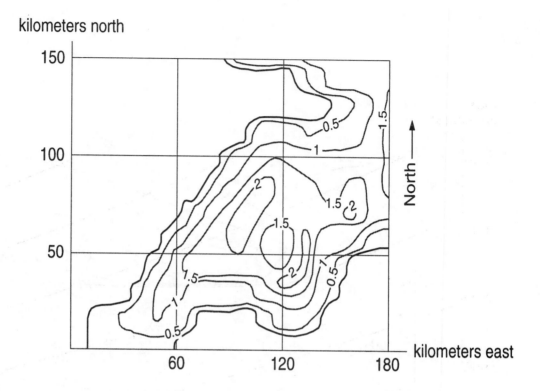

Transparency Master for Figure 7.40 for Problem 28, Section 7.2

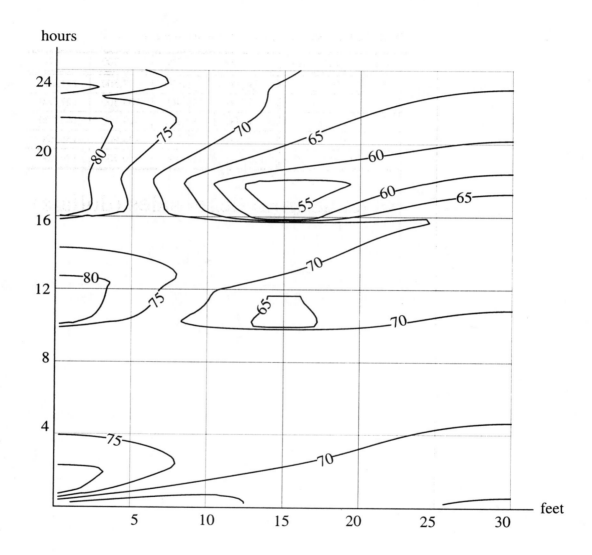

Transparency Master for Table 7.16 in Section 7.3

TABLE 0.0.3 *Revenue from ticket sales (dollars)*

Discount \ Full-Price	100	200	300	400
200	39,700	63,600	87,500	111,400
400	55,500	79,400	103,300	127,200
600	71,300	95,200	119,100	143,000
800	87,100	111,000	134,900	158,800
1000	102,900	126,800	150,700	174,600

Revenue from ticket sales (dollars)

PART V

CALCULATOR PROGRAMS

CALCULATOR PROGRAMS

for the

TI - 81

Program to Calculate Riemann Sums to Evaluate a Definite Integral (TI-81)

Select 'PRGM' to get the program menu, move to 'EDIT' to enter a program. When you select a program number, you must first give it a name (for example, 'RSUMS') to the right of the program number.

Prgm: RSUMS	*Where to Find The Commands*
:Disp "LOWER LIMIT"	Disp and Input are accessed via PRGM, I/O, pressed while editing a program.
:Input A	The " " are on the + key when you have pressed the ALPHA key.
:Disp "UPPER LIMIT"	
:Input B	
:Disp "DIVNS"	
:Input N	
:$A \to X$	$\to$ is on the STO key.
:$0 \to S$	0 is Zero, not oh.
:$1 \to I$	
:$(B - A)/N \to H$	
:Lbl P	Lbl is accessed via PRGM, CTL.
:$S + H * Y_1 \to S$	Y_1 is accessed via Y-VARS, i.e., 2nd VARS. (Do not use Y followed by 1, which means Y multiplied by 1.)
:$X + H \to X$	
:$IS > (I, N)$	$IS > ($ and Goto are accessed via PRGM, CTL.
:Goto P	
:Disp "LEFT SUM"	
:Disp S	
:$S + Y_1 * H \to S$	
:$A \to X$	
:$S - Y_1 * H \to S$	
:Disp "RIGHT SUM"	
:Disp S	

Things to watch for:

1. Difference between $-$, which means subtract, and (-), which means negative. For example, $-2 - 3$ must be entered as (-)2 $- 3$.

2. Disp and Input are not typed in letter-by-letter, but are obtained by highlighting them under PRGM, I/O and hitting ENTER.

To run this program:

1. The integrand (the function you want to integrate) $f(x)$ must be entered into Y_1, with X as the independent variable.

2. Make sure the lower limit of the integral you're approximating is less than the upper limit.

3. Test on $\displaystyle\int_1^3 x^3\,dx = 20$ with 100 subintervals. You should get left- and right-hand sums of 19.7408 and 20.2608, respectively.

Numerical Integration Program (TI-81)

This program calculates left- and right-hand Riemann sums, and the trapezoidal, midpoint and Simpson approximations. Since there's not room on the calculator to label each approximation separately, we use a compressed method of displaying the results. For instance, the label LEFT/RIGHT indicates that the next two numbers are the left- and right-hand Riemann sums, respectively.

Notes:

1. To enter a program, hit PRGM and select EDIT and a program number. To finish a line, hit ENTER; to finish editing a program, hit 2nd QUIT.

2. The function to be integrated must be entered as Y_1 (accessed by the "$Y =$" button). When Y_1 occurs in a program, it is evaluated at the current value of X.

3. The lower limit of integration must be less than the upper limit.

4. $IS > ($ means that the PRGM button must be pushed and then $IS > ($ selected, not that $I, S, >$ and (are to be entered separately. 'Disp' and 'Input' are to be found under PRGM, I/O, pushed while entering a program.

5. To run a program, select PRGM, EXEC. To stop a program while it is running, hit ON.

6. Test the program by evaluating $\int_1^3 x^3\, dx = 20$, using 100 subdivisions. You should get left- and right-hand sums of 19.7408 and 20.2608, respectively. For the trapezoid approximation, you should get 20.0008. For the mid point approximation, you should get 19.9996. For Simpson's rule, you should get exactly 20.

82

Prgm: INTEGRAL	*Where to Find The Commands*
:Disp "LOWER LIMIT"	Disp and Input are accessed via PRGM, I/O
:Input A	Enter lower limit of integration.
:Disp "UPPER LIMIT"	
:Input B	Enter upper limit of integration.
:Disp "DIVNS"	
:Input N	Enter number subdivisions.
:$(B - A)/N \rightarrow H$	Stores size of one subdivision in H (Note that $\rightarrow$ means hit STO button).
:$A \rightarrow X$	Start X off at beginning of interval.
:$0 \rightarrow L$	Initialize L, which keeps track of left sums, to zero.
:$0 \rightarrow M$	Initialize M, which keeps track of midpoint sums, to zero.
:$1 \rightarrow I$	Initialize I, the counter for the loop.
:Lbl P	Label for top of loop. Lbl is accessed via PRGM, CTL.
:$L + H * Y_1 \rightarrow L$	Increment L by $Y_1 H$, the area of one more rectangle. (Y_1 is accessed via Y-VARS, or 2nd VARS.)
:$X + .5H \rightarrow X$	Move X to middle of interval.
:$M + H * Y_1 \rightarrow M$	Evaluate Y_1 at the middle of interval and increment M by rectangle of this height.
:$X + .5H \rightarrow X$	Move X to start of next interval.
:$IS > (I, N)$	$IS > ($ is accessed via PRGM, CTL. This is the most difficult step in the program: adds 1 to I and does the the next step if $I \leq N$ (i.e., if haven't gone through loop enought times); otherwise, skips next step. Thus, if $I \leq N$, goes back to Lbl P and loops through again. If $I > N$, loop is finished and goes on to print out results. Continue here if $I > N$, in which case the value of X is now B.
:Goto P	Goto is accessed via PRGM, CTL. Jumps back to Lbl P if $J \leq N$.
:Disp "LEFT/RIGHT"	
:Disp L	L now equals the left sum, so display it.
:$L + H * Y_1 \rightarrow R$	Add on area of right-most rectangle, store in R.
:$A \rightarrow X$	Reset X to A.
:$R - H * Y_1 \rightarrow R$	Subtract off area of left-most rectangle.
:Disp R	R now equals right sum, so display it.
:$(L + R)/2 \rightarrow T$	Trap approximation is average of L and R.
:Disp "TRAP/MID/SIMP"	
:Disp T	Display trap approximation.
:Disp M	Display midpoint approximation.
:$(2M + T)/3 \rightarrow S$	Simpson is weighted average of M and T.
:Disp S	Display Simpson's approximation.

CALCULATOR PROGRAMS

for the

TI - 82

Introduction to the TI-82

The following are programs for the TI-82 calculator to do numerical integration, graph slope fields, demonstrate solutions to differential equations using Euler's method, and plot trajectories and solutions to systems of differential equations.

Notes on Using the TI-82: The commands on the TI-82 are basically the same as on the TI-81, though they may be in different menus. Select 'PRGM' to get the program menu, move to 'NEW' to enter a new program. You must first give it a name (for example, 'RSUMS') when prompted; to finish entering/editing a program, hit 2nd QUIT. To delete a program, press 2nd MEM and select [2:Delete]. The quantity AB represents the product of A and B. Writing $Y_3 = Y_1 Y_2$ defines Y_3 as the product Y_1 and Y_2. Writing $Y_3 = Y_1(Y_2)$ defines Y_3 as the composition of Y_1 and Y_2.

Program to Calculate Riemann Sums to Evaluate a Definite Integral (TI-82)

Select 'PRGM' to get the program menu, move to 'NEW' to enter a new program. You must first give it a name (for example, 'RSUMS') when prompted; to finish entering/editing a program, hit 2nd QUIT.

Name= RSUMS	*Where to Find The Commands*
:Disp "LOWER LIMIT"	Disp and Input are accessed via PRGM, I/O, pressed while in the middle of a program.
:Input A	The " " are on the + key when you have pressed the ALPHA key.
:Disp "UPPER LIMIT"	
:Input B	
:Disp "DIVNS"	
:Input N	
:$A \to X$	$\to$ is on the STO key.
:$0 \to S$	0 is Zero, not oh.
:$1 \to I$	
:$(B - A)/N \to H$	
:Lbl P	Lbl is accessed via PRGM, CTL.
:$S + H * Y_1 \to S$	Y_1 is accessed via Y-VARS, i.e., 2nd VARS. (Do not use Y followed by 1, which means Y multiplied by 1.)
:$X + H \to X$	
:$IS > (I, N)$	$IS > ($ and Goto are accessed via PRGM, CTL.
:Goto P	
:Disp "LEFT SUM"	
:Disp S	
:$S + Y_1 * H \to S$	
:$A \to X$	
:$S - Y_1 * H \to S$	
:Disp "RIGHT SUM"	
:Disp S	

Things to watch for:

1. Difference between $-$, which means subtract, and (-), which means negative. For example, $-2 - 3$ must be entered as (-)2 $-$ 3.
2. Disp and Input are not typed in letter-by-letter, but are obtained by highlighting them under PRGM, I/O and hitting ENTER.

To run this program:

1. The integrand (the function you want to integrate) $f(x)$ must be entered into Y_1, with X as the independent variable.
2. Make sure the lower limit of the integral you're approximating is less than the upper limit.
3. Test on $\int_{1}^{3} x^3 \, dx = 20$ with 100 subintervals. You should get left- and right-hand sums of 19.7408 and 20.2608, respectively.

Numerical Integration Program (TI-82)

This program calculates left- and right-hand Riemann sums, and the trapezoidal, midpoint and Simpson approximations. Since there's not room on the calculator to label each approximation separately, we use a compressed method of displaying the results. For instance, the label LEFT/RIGHT indicates that the next two numbers are the left- and right-hand Riemann sums, respectively.

Notes:
1. Select 'PRGM' to get the program menu, move to 'NEW' to enter a new program. You must first give it a name (for example, 'INTEGRAL') when prompted; to finish entering/editing a program, hit 2nd QUIT.
2. The function to be integrated must be entered as Y_1 (accessed by the "$Y =$" button). When Y_1 occurs in a program, it is evaluated at the current value of X.
3. The lower limit of integration must be less than the upper limit.
4. $IS > ($ means that the PRGM button must be pushed and then $IS > ($ selected, not that $I, S, >$ and (are to be entered separately. 'Disp' and 'Input' are to be found under PRGM, I/O, pushed while entering a program.
5. To run a program, select PRGM, EXEC. To stop a program while it is running, hit ON.
6. Test the program by evaluating $\int_1^3 x^3\, dx = 20$, using 100 subdivisions. You should get left- and right-hand sums of 19.7408 and 20.2608, respectively. For the trapezoid approximation, you should get 20.0008. For the mid point approximation, you should get 19.9996. For Simpson's rule, you should get exactly 20.

Name= INTEGRAL	*Where to Find The Commands*
:Disp "LOWER LIMIT"	Disp and Input are accessed via PRGM, I/O
:Input A	Enter lower limit of integration.
:Disp "UPPER LIMIT"	
:Input B	Enter upper limit of integration.
:Disp "DIVNS"	
:Input N	Enter number subdivisions.
:$(B - A)/N \to H$	Stores size of one subdivision in H (Note that $\to$ means hit STO button).
:$A \to X$	Start X off at beginning of interval.
:$0 \to L$	Initialize L, which keeps track of left sums, to zero.
:$0 \to M$	Initialize M, which keeps track of midpoint sums, to zero.
:$1 \to I$	Initialize I, the counter for the loop.
:Lbl P	Label for top of loop. Lbl is accessed via PRGM, CTL.
:$L + H * Y_1 \to L$	Increment L by $Y_1 H$, the area of one more rectangle. (Y_1 is accessed via Y-VARS, or 2nd VARS.)
:$X + .5H \to X$	Move X to middle of interval.
:$M + H * Y_1 \to M$	Evaluate Y_1 at the middle of interval and increment M by rectangle of this height.
:$X + .5H \to X$	Move X to start of next interval.
:$IS > (I, N)$	$IS > ($ is accessed via PRGM, CTL. This is the most difficult step in the program: adds 1 to I and does the the next step if $I \leq N$ (i.e., if haven't gone through loop enought times); otherwise, skips next step. Thus, if $I \leq N$, goes back to Lbl P and loops through again. If $I > N$, loop is finished and goes on to print out results. Continue here if $I > N$, in which case the value of X is now B.
:Goto P	Goto is accessed via PRGM, CTL. Jumps back to Lbl P if $J \leq N$.
:Disp "LEFT/RIGHT"	
:Disp L	L now equals the left sum, so display it.
:$L + H * Y_1 \to R$	Add on area of right-most rectangle, store in R.
:$A \to X$	Reset X to A.
:$R - H * Y_1 \to R$	Subtract off area of left-most rectangle.
:Disp R	R now equals right sum, so display it.
:$(L + R)/2 \to T$	Trap approximation is average of L and R.
:Disp "TRAP/MID/SIMP"	
:Disp T	Display trap approximation.
:Disp M	Display midpoint approximation.
:$(2M + T)/3 \to S$	Simpson is weighted average of M and T.
:Disp S	Display Simpson's approximation.

CALCULATOR PROGRAMS

for the

TI - 85

Introduction to the TI-85

The following are programs for the TI-85 calculator to do numerical integration, graph slope fields, demonstrate solutions to differential equations using Euler's method, and plot trajectories and solutions to systems of differential equations.

Notes on Using the TI-85: When you want to multiply two quantities, say A and B, on the TI-85 you must use "∗". On a TI-81, AB means $A ∗ B$, whereas on a TI-85 AB is a single variable. If you can't find a command, try 2nd, CATALOG, which gives a list of all the commands. Press a letter to go quickly to the commands beginning with that letter. Press ENTER to select. Letters can be both upper and lower case. Upper case letters are obtained by hitting ALPHA; lower case by hitting 2nd ALPHA. To enter a function (for example, $y1 =$) under the GRAPH menu, lower case variables should be used.

Program to Calculate Riemann Sums to Evaluate a Definite Integral (TI-85)

Select 'PRGM' to get the program menu, then select 'EDIT' to enter a program. When you enter a program, you must first give it a name (for example, 'RSUMS'). To finish editing, hit EXIT.

	Where to Find The Commands
Name=RSUMS	
:Disp "LOWER LIMIT"	Disp, " ", and Input are accessed via I/O, pressed while in the middle of a program.
:Input A	
:Disp "UPPER LIMIT"	
:Input B	
:Disp "DIVNS"	
:Input N	
:$A \to x$	$\to$ is on the STO key; use x-VAR key for x.
:$0 \to S$	0 is Zero, not "oh".
:$1 \to I$	
:$(B - A)/N \to H$	
:Lbl P	Lbl is accessed via CTL or CATALOG.
:$S + y1 * H \to S$	$y1$ is typed in by entering 2nd ALPHA Y and then 1.
:$x + H \to x$	
:$IS > (I, N)$	$IS >$ (and Goto are accessed via CTL or CATALOG.)
:Goto P	
:Disp "LEFT SUM"	
:Disp S	
:$S + y1 * H \to S$	
:$A \to x$	
:$S - y1 * H \to S$	
:Disp "RIGHT SUM"	
:Disp S	

Things to watch for:

1. Difference between $-$, which means subtract, and (-), which means negative. For example, $-2 - 3$ must be entered as (-)2 $-$ 3.

2. Disp and Input can also be typed in letter-by-letter.

To run this program:

1. The integrand (the function you want to integrate) $f(x)$ must be entered into y_1, under the GRAPH menu, with x as the independent variable.

2. Make sure the lower limit of the integral you're approximating is less than the upper limit.

3. Test on $\int_1^3 x^3 \, dx = 20$ with 100 subintervals. You should get left- and right-hand sums of 19.7408 and 20.2608, respectively.

Numerical Integration Program (TI-85)

This program calculates left and right Riemann sums, and the trapezoidal and midpoint approximations. Since there's no room on the calculator for a separate labeling of each approximation, we use a compressed method of displaying the results. For instance, the label LEFT/RIGHT indicates that the next two numbers are the left- and right-hand Riemann sums, respectively.

Notes:

1. Select 'PRGM' to get the program menu, then select 'EDIT' to enter a program. When you enter a program, you must first give it a name (for example, 'INTEG'). To finish editing, hit EXIT.

2. The function to be integrated must be entered as $y1$ (accessed by GRAPH, followed by $y(x) =$). When $y1$ occurs in a program, it is evaluated at the current value of x.

3. The lower limit of integration must be less than the upper limit.

4. $IS > ($ is selected from under CTL, while enter entering the program. "Disp" and "Input" are selected from under I/O.

5. Use MORE to see items on a menu which are currently off the screen.

6. To run a program, select PRGM, NAMES. To stop a program while it is running, hit ON.

7. Test the program by evaluating $\int_1^3 x^3 \, dx = 20$, using 100 subdivisions. You should get left- and right-hand sums of 19.7408 and 20.2608, respectively. For the trapezoid approximation, you should get 20.0008. For the mid point approximation, you should get 19.9996. For Simpson's rule, you should get exactly 20.

Name=INTEG	*Where to Find The Commands*
:Disp "LOWER LIMIT"	Disp, Input, and " " are accessed via PRGM, I/O.
:Input A	Enter lower limit of integration.
:Disp "UPPER LIMIT"	
:Input B	Enter upper limit of integration.
:Disp "DIVNS"	
:Input N	Enter number subdivisions.
:$(B - A)/N \to H$	Stores size of one subdivision in H. (Note that $\to$ means hit STO button).
:$A \to x$	Start x off at beginning of interval.
:$0 \to L$	Initialize L, which keeps track of left sums, to zero.
:$0 \to M$	Initialize M, which keeps track of right sums, to zero.
:$1 \to I$	Initialize I, the counter for the loop.
:Lbl P	Label for top of loop. Lbl is accessed via CTL
:$L + H * y1 \to L$	Increment L by $H * y1$, the area of one more rectangle. ($y1$ is typed in as y and then 1.)
:$x + 0.5H \to x$	Move x to middle of interval.
:$M + H * y1 \to M$	Evaluate $y1$ at the middle of interval and increment M by a rectangle of this height.
:$x + 0.5H \to x$	Move x to start of next interval.
:$IS > (I, N)$	Access $IS > ($ from under CTL. This is the most difficult step in the program: adds 1 to I and does the next step if $I \leq N$ (i.e., if haven't gone through loop enought times); otherwise, skips next step. Thus, if $I \leq N$, goes back to Lbl 1 and loops through again. If $I > N$, loop is finished and goes on to print out results.
:Goto P	Goto is accessed via CTL. Jumps back to Lbl P if $I \leq N$.
:Disp "LEFT/RIGHT"	
:Disp L	L now equals the left sum, so display it.
:$L + H * y1 \to R$	Add on area of right-most rectangle, store in R.
:$A \to x$	Reset x to A.
:$R - H * y1 \to R$	Subtract off area of left-most rectangle.
:Disp R	R now equals right sum, so display it.
:$(L + R)/2 \to T$	Trap approximation is average of L and R.
:Disp "TRAP/MID/SIMP"	
:Disp T	Display trap approximation.
:Disp M	Display midpoint approximation.
:$(2 * M + T)/3 \to S$	Simpson is weighted average of M and T.
:Disp S	Display Simpson's approximation.

CALCULATOR PROGRAMS

for the

CASIO fx-7700GB

Introduction to the CASIO fx-7700GB

The following are programs for the CASIO fx-7700GB calculator to do numerical integration, graph slope fields, demonstrate solutions to differential equations using Euler's method, and plot trajectories and solutions to systems of differential equations.

Notes on Using the CASIO fx-7700GB:

- All the Casio integration programs call on the function you put in f_1. To store a function in f_1, first select the Function Memory Menu by pressing SHIFT then $\boxed{F}$MEM. Clear the screen by pressing AC if needed. Type out the function, then press STO (F1 key) followed by 1. Press LIST (F4 key) to see the list of stored functions.

- You can put either a colon (:) or a carriage return (EXE) after each instruction (to separate them), except after display sign▶ , which provides its own carriage return.

Riemann Sums Program (CASIO)

To enter the program, press 'MODE' then '2' to select 'WRT' mode. Move the cursor to an empty program number, then press 'EXE'. You'll see a blank screen with the blinking cursor at the upper left corner. Now you can proceed to the beginning of the program. When finished, press 'MODE' then '1' to get back to 'RUN' mode.

Program	*Where to Find The Commands*
"RSUMS"	
"L-LIM" ?$\rightarrow$ A	'?' is accessed by SHIFT, PRGM
"R-LIM" ?$\rightarrow$ B	
"DIVNS" ?$\rightarrow$ N	
A $\rightarrow$ X	
0 $\rightarrow$ S	
$f_1 \rightarrow$ Y	'f_1' is accessed by SHIFT, $\boxed{F}$MEM, f_n, 1
(B$-$A)div N $\rightarrow$ H	
Lbl 1	'Lbl' is in JMP menu
X + H $\rightarrow$ X	
f_1 + S $\rightarrow$ S	
Dsz N	'Dsz' is in JMP menu
Goto 1	'Goto is also in JMP menu
"L-SUM="	
(S $-$ f_1 + Y)H$\blacktriangleright$	
"R-SUM="	
HS $\blacktriangleright$	

To run this program:

1. The integrand (the function you want to integrate) $f(x)$ must be entered into f_1, with X as the independent variable.

2. Make sure the lower limit of the integral you're approximating is less than the upper limit.

3. Test on $\int_1^3 x^3\, dx = 20$ with 100 subintervals. You should get left- and right-hand sums of 19.7408 and 20.2608, respectively.

Numerical Integration Program (CASIO)

This is a **Casio fx series** calculator program for various numerical integrals. It will display the Left and Right Riemann Sums and the Trapezoid Rule, Midpoint Rule, and Simpson's Rule approximations all at once. The way this program evaluates integrals is by keeping a running total of function values on n subintervals, and then multiplying by the width of the rectangles to obtain the area at the very end. At the end of the program, hitting EXE will let you reevaluate the integral with a different number of subdivisions, and hitting AC will let you out of the program.

Program	Comments
"INTEGRAL"	
"L-LIM"?→A	Integrate from X=a
"U-LIM"?→B	to X=b
"DIVNS"?→N	over N subdivisions.
(B−A)div (2N)→H	calculates half the width of a subdivision.
0→L	initialize L, which will keep track of the left sums.
0→M	initialize M, the midpoint sum.
A→X	place X at A, the beginning of the interval.
Lbl 1	top of loop:
f_1+L→L	evaluate the function at the left edge of the interval add the result to the left-hand sum running total.
X+H→X	move X to the middle of the interval.
f_1+M→M	evaluate the function at the middle of the interval add the result to the midpoint running total.
X+H→X	move X to the beginning of the next interval.
Dsz N	decrease N by 1; if N=0, skip the next step and go on.
Goto 1	bottom of loop.
"LEFT,RIGHT,TRAP"	
2HL→L	multiply the sum of the left-hand function values by width 2H.
L▸	display the left-hand sum.
L+2Hf_1→T	evaluate the function at X=b: the rightmost function value add the area of the rightmost rectangle with the left-hand sum.
A→X	put X back at A.
T−2Hf_1→T	evaluate the function at the left-most edge of the interval take the area of the leftmost rectangle out of T.
T▸	display what is now the right-hand sum.
(L+T)div 2→T	average the left- and right-hand sums.
T▸	display the trapezoid approximation.
"MID,SIMP"	
2HM→M	multiply the midpoint values sum by the interval width.
M▸	display the midpoint approximation.
(2M+T)div 3→M	calculate Simpson's Rule by weighted averaging.
M	display Simpson's Rule approximation.

CALCULATOR PROGRAMS

for the

SHARP EL-9200 and EL-9300

Introduction to the Sharp EL-9200 and EL-9300

The following programs are for the Sharp EL-9200 and Sharp EL-9300 calculators. Enclosed you will find programs to do numerical integration, graph slope fields, demonstrate solutions to differential equations using Euler's method, and plot trajectories and solutions to systems of differential equations.

Notes:

1. All the following programs are in REAL mode.

2. To access the commands, first press 2ndF COMMAND, then select them via the appropriate menus.

3. After editing a line in a program, make sure to press ENTER or the button with downward pointing triangle before you quit; otherwise the changes will not be saved.

4. Since the user-defined functions (Y_1 ... etc.) are not shared by different programs, the formula for a function has to be given when it is used in a program. Thus, each programs has a subroutine that starts with the line "Label eqn" and ends with "Return", where the desired function should be entered.

5. Variables are case sensitive. Single uppercase letters (A to Z) are global variables, i.e., the values stored in memories designated by single letter A to Z can be shared by different programs. Lowercase letters and lowercase words are local variables, i.e., the values of local variables are specific to the program in which they're used. You can use a string of up to 12 lowercase letters to designate a local variable. Note also that you can not mix uppercase and lowercase letters to form a variable.

Riemann Sums Program (SHARP)

Select NEW in the program menu. Then select REAL in the MODE menu. When prompted for a title, use "riemann." The "..." in the following program is where the integrand–the function to be integrated–should be entered.

Program	Where the Commands are
Program	*Where the Commands are*
Goto start	Goto is in BRANCH menu
Label eqn	Label is in BRANCH menu
f=...	replace "..." by the integrand $f(x)$
Return	Return is in BRANCH menu
Label start	
Print "l-limit	Print and " are in PROG menu
Input a	Input is in PROG menu
Print "u-limit	
Input b	
Print "divns	
Input n	
x=a	= is also in INEQ menu
s=0	reset memory s to zero
i=1	
h=(b−a)/n	
Label 1	
Gosub eqn	Gosub is in BRANCH menu
s=s+f*h	
x=x+h	
i=i+1	
If i <= n Goto 1	If and Goto are in BRANCH menu; <= is in INEQ menu
Print "left sum	
Print s	
Gosub eqn	
s=s+f*h	
x=a	
Gosub eqn	
s=s−f*h	
Print "right sum	
Print s	
End	End is in PROG menu

To run this program:

1. The integrand (the function you want to integrate) $f(x)$ must be entered into where "..." is, with x as the independent variable.

2. Make sure the lower limit of the integral you're approximating is less than the upper limit.

3. Test on $\int_1^3 x^3\,dx = 20$ with 100 subintervals. You should get left- and right-hand sums of 19.7408 and 20.2608, respectively.

Numerical Integration Program (SHARP)

This program calculates left- and right-hand Riemann sums, and the trapezoidal, midpoint and Simpson approximations. Since there's not room on the calculator to label each approximation separately, we use a compressed method of displaying the results. For instance, the label "left/right" indicates that the next two numbers are the left- and right-hand Riemann sums, respectively.

To enter the program select NEW in the program menu. Then select REAL in the MODE menu. When prompted for a title, use "integral."

Program	Where the Commands are
Goto start	Goto is in BRANCH menu
Label eqn	Label is in BRANCH menu
f=...	replace "..." by the integrand $f(x)$
Return	Return is in BRANCH menu
Label start	
Print "l-limit	Print and " are in PROG menu
Input a	Input is in PROG menu
Print "u-limit	
Input b	
Print "divns	
Input n	
x=a	= is also in INEQ menu
s=0	reset memory s to zero
m=0	reset memory m to zero
i=1	
h=(b−a)/n	
Label 1	
Gosub eqn	Gosub is in BRANCH menu
s=s+f*h	
x=x+ .5h	
Gosub eqn	
m=m+f*h	
x=x+ .5h	
i=i+1	
If i <= n Goto 1	If and Goto are in BRANCH menu; <= is in INEQ menu
Print "left/right	
Print s	
Gosub eqn	
r=s+f*h	
x=a	
Gosub eqn	
r=r−f*h	
Print r	
Wait	Wait is in PROG menu
Print "trap/mid/simp	
t=.5(s+r)	
Print t	
Print m	
s=(2m+t)/3	
Print s	
End	End is in PROG menu

To run this program:

1. The integrand (the function you want to integrate) $f(x)$ must be entered into where "..." is, with x as the independent variable.

2. Make sure the lower limit of the integral you're approximating is less than the upper limit.

3. Test on $\int_1^3 x^3\, dx = 20$ with 100 subintervals. You should get left- and right-hand sums of 19.7408 and 20.2608, respectively.

CALCULATOR PROGRAMS

for the

HP-48S, HP-48G and HP-38G

Riemann Sums for the HP-48S/G

The following is a directory for experimenting with different Riemann sums. To use a directory press VAR to get the user's menu and then press the name of directory on menu keys, in this case RSUM. To leave the directory when you are finished, press (left shift) UP. To create a directory, type the name, say 'RSUM' followed by CRDIR on the MEMORY menu.

RSUM Directory

These short programs can be entered by hand or transferred via infrared from another HP-48. Programs are given in the order you will find most convenient to use on the menu. The name of each program is given before the program. As you enter each program, store it under the given name. Thus for first program, type << ANS − >> followed by ENTER and then type 'ERR' and press STO.

ERR << ANS − >>
LFT << A SUM >>
RGT << A H + SUM >>
MID << A H 2 / + SUM >>
TRP << LFT B F A F − H * 2 / + >>
SMP << MID 2 * TRP + 3 / >>
NSTO << 'N' STO B A − N / 'H' STO >>
ABSTO << 'B' STO 'A' STO >>
FSTO << 'F(X)' SWAP = DEFINE >>
SUM << → X 'H*Σ(I=0, N−1,F(X+I*H))' >>

The other variables used by these programs will also be on your menu: A, B, F, N, H, ANS. To make sure your menu is in the most convenient order, just use FSTO, ABSTO, and NSTO once, which will creat variables A, B, F, N, and H. Also do 0 'ANS' STO to create a variable called ANS. Then press (left shift) {} and press menu buttons to get this:
{ERR LFT RGT MID TRP SMP NSTO ABSTO FSTO ANS}
Press ENTER. Then type in ORDER followed by ENTER (or find ORDER on the MEMORY menu and press the menu button).

How to Use Directory RSUM

Here is an example. If you want to experiment with $\int_0^1 \frac{1}{1+x^2}\,dx$ with $N=10$ subdivisions do this:
Enter '1/(1 + X^2)' and press FSTO.
Enter 0. Enter 1. Press ABSTO.
Enter 10 and press NSTO.
Now press LFT, RGT, MID, TRP, SMP to get the left, right, midpoint, trapezoid, and Simpson approximations for the given integral. If you know the actual value of the integral, store that value in ANS. Then when you press ERR the value of ANS will be subtracted from whatever is at level one of the stack. Thus, if you press MID followed by ERR, the error for the midpoint approximation will appear on the stack. This is especially useful if you are trying to study how the errors of the different methods are related to each other and to N.

Riemann Integration on the HP 38G

Enter the programming environment on the HP 38G by pressing (shift) PROGRAM. There you will see a list of all programs in the machine, plus the name Editline. To enter a new program, press [NEW] on the menu bar. You will be prompted for a name for the program; we call this one R.INT. Type in R.INT and press [ENTER]. A screen titled R.INT PROGRAM will appear. Type in the following program; you may look for commands on the menus, but it is far easier to type the letters one by one. To type a string of capital letters, hold down the [**A...Z**] button and type letters; lower case letters must be typed with (shift) a...z before each one. The symbol ▶ is typed by pressing [**STO ▶**] on the menu. Other special symbols (", =, ?) are typed from the Character Browser, accessed by typing (shift) CHARS. To get a new line, press [**ENTER**]. Indentation is done here for ease in reading; on the machine, it is optional.

Program **R.INT**:

```
SELECT Function:                    0 ▶ L:
2 ▶ Format:                         FOR I=1 TO N STEP 1;
9 ▶ Digits:                             L+F1(X) ▶ L:
MSGBOX                                  X+H ▶ X:
    "STORE INTEGRAND IN F1":        END:
INPUT A;                            L+F1(X) ▶ R:
    "LOWER LIMIT";                  L*H ▶ L:
    "A";                            A ▶ X:
    "ENTER LOWER LIMIT";            R-F1(X) ▶ R:
    0:                              R*H ▶ R:
INPUT B;                            MSGBOX
    "UPPER LIMIT;                       "LHS=" L
    "B";                                "RHS=" R
    "ENTER UPPER LIMIT";                "AVE=" (L+R)/2:
    1:                              1 ▶ C:
DO                                  CHOOSE C;
    INPUT N;                            "DO IT AGAIN?";
        "SUBINTERVALS";                 "YES";
        "N;                             "NO":
        "ENTER NUMBER OF            UNTIL
SUBINTERVALS";                          C==2
    10:                             END:
    (B-A)/N ▶ H:                    1 ▶ Format:
    A ▶ X:
```

The program is saved automatically. To put the integrand in F1, press [**LIB**] and highlight Function; then press [**START**] on the menu bar, highlight F1, and press [**EDIT**], then type in the formula. To run the program, press (shift) PROGRAM and press [**RUN**].

PART VI

SAMPLE EXAM QUESTIONS

1. For a function $Q = f(r)$, the average rate of change over the interval from $r = r_0$ to $r = r_1$ is given by what?

2. The formula $P = 23.6(0.975)^t$ describes the size of a population (in millions) with t measured in years. What can you conclude about the population simply by looking at the formula?

3. If f is a function, what does the following expression represent?

$$\frac{f(2) - f(1.999)}{0.001}$$

4. What is the doubling time of the function $P = 20(4)^t$?

5. Graph the following situation: (Clearly indicate what each of the axes represents). "Fertilisation increases the yield of a harvest, but over fertilisation proves to be detrimental to the yield."

6. In South Africa, as in many other countries, using a TV requires a license. An advertising campaign is launched in order to encourage viewers to pay their TV licenses. Table 0.2 depicts the number of paying viewers (in millions) against the advertising cost (in millions of rands). A rand, R, is the South African unit of currency.

TABLE 0.1

Cost of the campaign	5.3	12.5	18.2	24.5
Number of paying viewers	2.3	2.8	3.1	3.3

(a) Find the average rate of change at which the number of viewers change against the advertising cost, using the last two inscriptions in the table.

(b) Interpret your answer.

(c) Suppose the licence fee is R200 per viewer. Why should (shouldn't) more be spent on the campaign?

7. Study the supply and demand curves in Figure 0.1. Discuss the supply and demand situation if the price is to be R500/item. what will the probable consequence be?

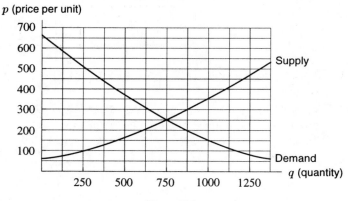

Figure 0.1

8. Study the following cost and revenue table. Find the cost function C as well as the revenue function R and use these functions to determine the number of items that have to be sold in order to make a profit.

TABLE 0.2

q	100	200	300	400	500
$C(q)$	3500	4000	4500	5000	5500
$R(q)$	800	1600	2400	3200	4000

9.   It takes two minutes for the temperature of a cup of coffee to decrease from 96°C to 72°C.

   (a)   If the temperature decreases linearly, then the temperature of the coffee will be
   _____ after four minutes.

   (b)   Of the temperature decreases exponentially, then the temperature of the coffee will be
   _____ after four minutes.

   (c)   Find a formula that describes the exponential case.

   (d)   Use the information given above as well as your answers in (a) and (b) to show that the absolute rate of change is constant in the linear case and that the relative rate of change is constant in the exponential case.

10.   (a)   The graph of $y = x^2$ is given in Figure 0.2. Add to this the graphs of $y = x^{4/3}$ and $y = x^4$. Label them clearly and give the values of the points of intersection.

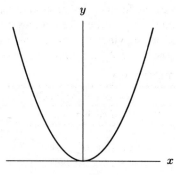

Figure 0.2

   (b)   Which of the curves $y = at^2$, $y = at^3$, or $y = a^t$ describes the following set of data best? Justify your choice.

TABLE 0.3

t	2	2.5	3	3.5
y	2.4	3.6	8.1	12.8

11.   Consider the function $f(t) = 2.6(1.12)^t$. If $f'(1.5) \approx 0.33215$ correct to the second decimal? Motivate clearly.

12.   Study the graph of the function f that shows the number of daylight hours in Washington for a six month period from 21 December. Estimate $f'(3)$ (units included).

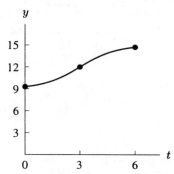

Figure 0.3: Daylight (in hours) as
a function of number of months
after December 21

13. (a) $7 \cdot 3^t = 5 \cdot 2^t$ $t =$ _____

    (b) $8 \cdot (2.5^x) = a \cdot e^{kx}$ $a =$ _____ $k =$ _____

14. The population of Nicaragua was 3.6 million in 1990 and growing at 3.4% per year. Let P be the population in millions, and let t be the time in years since 1990.

    (a) Express P as a function in the form $P = P_0 a^t$.

    (b) Express P as an exponential function using base e.

    (c) How long does it take for the population of Nicaragua to increase by 50%?

15. Find the equation of the line.

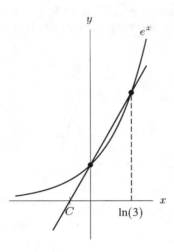

16. Find a linear approximation for the function $f(x) = \dfrac{2x + 8}{x - 2}$ valid for x near 3.

    $\dfrac{2x + 8}{x - 2} \approx$ _____ for x near 3.

17. The graph of $f(x)$ is given below.

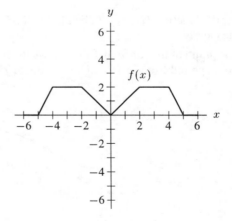

    Sketch the graphs of:

    (a) $2 + f(x)$ (b) $2f(x)$ (c) $1 - f(x)$ (d) $\dfrac{1}{f(x)}$

(a)
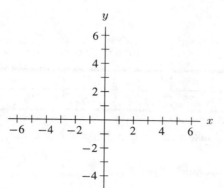

(b)
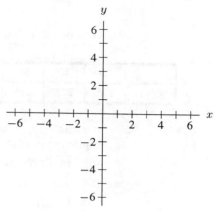

(c)
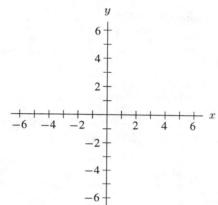

(d)
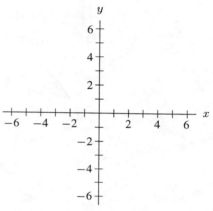

18. Find a possible formula for the following graphs:

(a)
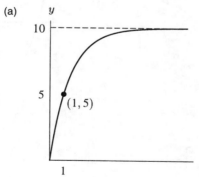

(b)
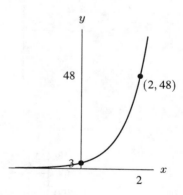

(c)
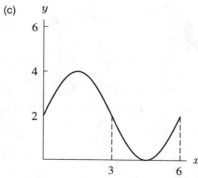

(d)
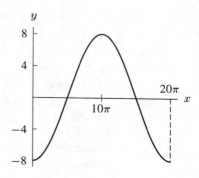

114

19. Give possible formulas for f and g.

x	20	40	60	80	100
$f(x)$	175	225	275	325	375

x	6	6.2	6.4	6.6	6.8
$g(x)$	14.93	15.48	16.06	16.66	17.27

20. A clean up of a polluted lake will remove 3% of the remaining contaminants every year, beginning in 1996.

    (a) In which year will the most be removed? _____

    (b) The goal is to reduce the quantity of contaminants to $\frac{1}{10}$ its present level. When will this be achieved? _____

21. Sketch $g'(x)$ and $f(x)$, assuming $f(0) = 0$.

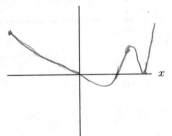

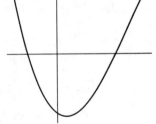

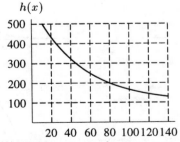

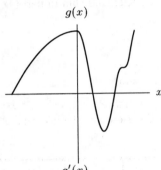

22.

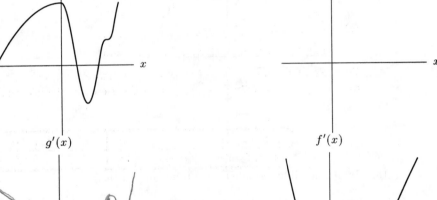

$h'(80) \approx$ _____

23. Using a difference quotient, compute $f'(1) \approx$ _____ for $f(x) = \sin(3x)$.

24. Let $f(T)$ be the time, in minutes, that it takes for an oven to heat up to $T°F$. What is/are the
    units of $f'(T)$ _____
    sign of $f'(T)$ _____
    meaning of $f(300) = 10$
    meaning of $f'(300) = 0.1$.

25.

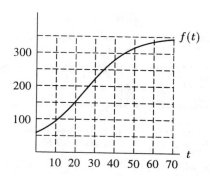

(a) $\displaystyle\int_{10}^{70} f(t)\,dt \approx$ _____

(give three terms of the left Riemann sum).

(b) Shade the area you computed with the Riemann sum in part (a). Indicate on the graph the error in your estimate of $\displaystyle\int_{10}^{70} f(t)\,dt$.

(c) $\displaystyle\int_{10}^{70} f(t)\,dt \approx$ _____

(give three terms of the right Riemann sum).

26. The rate of pollution pouring into a lake is measured every 10 days, with results in the following table. About how much pollution has entered the lake in the first 40 days?

Time in days	0	10	20	30	40
Rate of pollution (tons/day)	5	7	10	9	8

27. The number of hours, H, of daylight in Madrid as a function of the date is given by the formula

$$H = 12 + 2.4\sin(0.0172(t-80))$$

where t is the number of days since the beginning of the year.

(a) What are the units of $\dfrac{dH}{dt}$? _____

(b) Explain in words the meaning of $\left.\dfrac{dH}{dt}\right|_{t=100}$.

(c) Estimate $\left.\dfrac{dH}{dt}\right|_{t=100} \approx$ _____.

28. This table gives the wind chill factor (°F) as a function of wind speed (miles/hour) when air temperature is 20° F.

Wind speed (mph)	5	10	15	20	25
Wind chill factor (°F)	16	3	−5	−10	−15

(a) Give an approximation (including units) of the derivative of wind chill with respect to wind speed when the air temperature is 20°F and the wind speed is 10 miles/hour.

(b) Explain in practical every day language the meaning of the number you computed in part(a).

29.   Approximate (with difference quotient and calculator) the derivative of $\sqrt{8x+1}$ at $x = 1$.

30.   Indicate a scale on the axes for the graph of the derivative and sketch the graph of $f'(x)$.

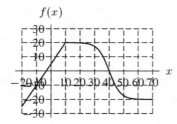   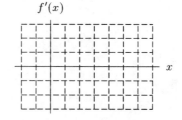

31.   Give a three term left Riemann sum approximation for the definite integral

$$\int_0^{12} \sqrt{8x+1}\,dx \approx \underline{\hspace{4cm}}$$

32.   The average value of $f(x)$ on the interval $0 \le x \le 100$ equals $\underline{\hspace{3cm}}$.

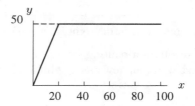

33.   If $g'(x) > 0$ then g is $\underline{\hspace{4cm}}$.
      If $g'(x)$ is decreasing then g is $\underline{\hspace{4cm}}$.
      If $g''(x) < 0$ then $g'(x)$ is $\underline{\hspace{4cm}}$.
      If $g''(x) < 0$ then g is $\underline{\hspace{4cm}}$.

34.   A mudslide in California is pouring mud into a valley at a rate of $h(t)$ ft^3/hour, where t is the number of hours since the slide began. We have $\int_0^{20} h(t)\,dt = 2{,}000{,}000$. Explain in practical every day language the meaning of this equation.

35.   Sketch the graph of $g(x)$, assuming $g(0) = 0$.

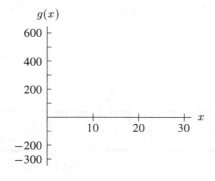   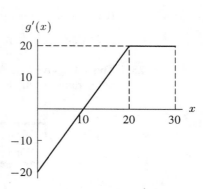

36.   Compute the derivative in all cases. Do not simplify your answer.

      (a) $y = 2x^2 + 5^x$
      (b) $y = x^2 \cos(7x + 5)$

(c)　$y = 10^{-x} - \tan^{-1}(x)$

(d)　$y = \dfrac{(8x + 1)}{(x^2 + 5)}$

(e)　$y = \ln(5x^4)$

(f)　$y = e^{\ln(x)}$

(g)　$y = e^{\ln(2)}$

(h)　$y = \sin(\cos(3x))$

37.　Find the best linear approximation for $f(x) = \dfrac{(8x + 1)}{(x^2 + 5)}$ valid for x near 0. (See Problem 36(d).)

$f(x) = \dfrac{(8x + 1)}{(x^2 + 5)} \approx$ _____ for x near 0.

38.　The depth of the water in feet of February 10, 1990 at a certain point in Boston harbor was given by the function $f(x) = 5 + 4.9\cos(0.5t - 1)$, where t is the number of hours since midnight.

(a)　What is the meaning of $f(3)$?

(b)　Was the tide rising or falling at 6am?

(c)　At what rate (with units) was it rising or falling at 6am?

39.　(a)　Find a function $f(x)$ such that $f'(x) = x^2 + 5x$.

　　　$f(x) =$ _____

(b)　Find the point at which the curve $y = x^2$ has slope equal to 10.

40.　Give possible formulas for f, g, and h.

(a)

x	11	13	15	17	19
$f(x)$	179.2	233	302.8	393.7	511.8

(b)

x	11	13	15	17	19
$g(x)$	100	104	108	112	116

(c)

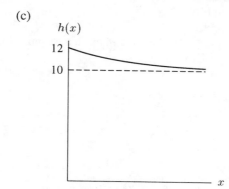

41.　The population of a certain country is growing exponentially. In 1980 the poplation was 3.5 million and in 1990 it was 4 million. Assume that exponential growth continues well to the future.

(a)　What population would you predict for the year 2010?

(b)　What population would you predict for t years from 1980?

(c)　What is the doubling time for the population of this country?

42. (a) Using a difference quotient, give an approximation for the derivative of $(x - 1)^{2x}$ at $x = 3$. Show the difference quotient you use.

    (b) Give (approximately) the equation of the line tangent to the graph of $f(x) = (x - 1)^{2x}$ at the point where $x = 3$.

43. Let $f(x)$ equal the weekly rent (in \$) of a holiday apartment in Gulf Shores at a distance of x feet from the shore line. Explain, in practical terms, the meaning of the statement $f'(200) = -0.3$.

44. (a) Sketch the graph of the derivative of $f(x)$, given the graph of $f(x)$.

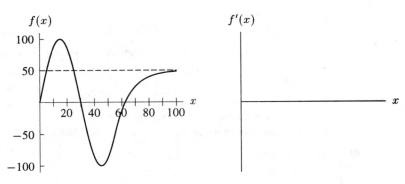

    (b) Estimate the following:
    $f'(10) = $ _____
    $f'(30) = $ _____

45. Approximate $\displaystyle\int_{40}^{60} \sqrt{x + 5}\, dx$ with a four term Riemann sum. Show all the terms in the sum.

46. Let $f(t)$ equal the rate (degrees Fahrenheit/minute) at which a pot of water is heating up at time t minutes after being put on the stove. Explain, in practical terms, the meaning of the statement $\displaystyle\int_{3}^{5} f(t)\, dt = 24$.

47. The graph shows the derivative f' of a mystery function f.

    (a) What are the critical points of f?
    (b) Where are the inflection points of f?
    (c) On what interval(s) is f increasing?
    (d) For what x in the range $0 \le x \le 100$ does $f(x)$ have a maximum value?
    (e) You are told that $f(25) = 50$. Evaluate $f(50)$.

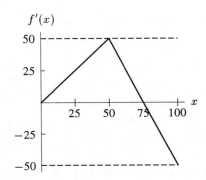

48. Compute the derivative in all cases. Do not simplify your answer.

(a) $\dfrac{5^x + \cos(x^2 + 5)}{1 + \sqrt{x}}$

(b) $y = e^{-x} \cos(7x + 5)$

(c) $y = \sin((x^2 + 1)^5)$

(d) $y = \ln(8x + 7)$

49. Give a linear approximation for the piece of the graph of $x^3 + xy + y^3 = 11$ near the point $(1, 2)$.

50. (a) Sketch the graph of $y = 100xe^{-x}$.

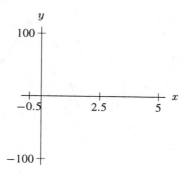

(b) Where is (are) the critical point(s)?

(c) Where is (are) the inflection point(s)?

(d) On what interval(s) is y increasing as a function of x?

(e) On what interval(s) is the graph concave up?

(f) What is the maximum value of y and for what value of x does it occur?

51. A lighthouse keeper wants to get from his lighthouse (10 miles from shore) to his house (10 miles down the beach from the point on the beach nearest the light house) in the shortest possible time. He will row (3 miles/hour) to the beach and then run (9 miles/hour) the rest of the way. What point on the beach should he row to?

52. The size of a bird population can be represented by a sinusoidal graph. The number decreased from a maximum of 20,000 in 1943 to a minimum of 12,000 in 1989 and then it increased again.

(a) Take 1943 as $t = 0$ and find a function that has this behavior.

(b) How many birds do you expect there to be on the island in the year 2000?

(c) For what single period from 1989 will the population show a positive growth?

53. An amount of $500 that was invested in 1970 increased as follows. (Amount given at the beginning of the year.)

TABLE 0.4

Year	1970	1975	1980	1985	1990	1995
Capital	500	966	1856	3578	6876	13233

(a) Find the average rate at which the capital increased
  (i) During the seventies
  (ii) During the nineties (using the information up to 1995)

(b) Compare the growth rate over both these intervals as percentages of the amounts at the beginning of each of the two periods and comment on this.

(c) Find an approximation to the instantaneous growth rate at the beginning of 1990.

(d) What additional information would you need in order to make a better approximation in part (c)?

54. Consider the function f in Figure 0.4. Decide whether each of the following is positive, negative or zero, and indicate each of these expressions graphically on the sketch above.

    (a) $\dfrac{f(4) - f(2)}{2}$

    (b) $\lim\limits_{h \to 0} \dfrac{f(2 + h) - f(2)}{h}$

    (c) $f''(3)$

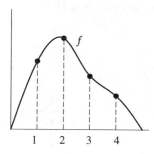

Figure 0.4

55. Find $f'(x)$ algebraically by using the limit definition if

$$f(x) = \frac{1}{x + 2}.$$

56. A factory manufactures T-shirts. The graph in Figure 0.5 shows the number of T-shirts, T, manufactured in 1 day against the number of workers, w, emplyed at the time. The factory can barely accomodate 50 workers.

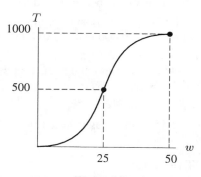

Figure 0.5

    (a) What are the units of $T'(w)$?
    (b) What is the physical meaning of $T'(20) = 10$?
    (c) What effect does the number of employed workers have on the productivity of the factory?
    (d) Sketch f'.
    (e) Sketch f''.
    (f) What does the derivative of $f^{-1}(20)$ tell you?

57. The internet is an electronic network allowing transfer of information. The sketch in Figure 0.6 gives the number, N, of messages (in billions) sent per month across the internet as a function of t, the number of years since 1990.

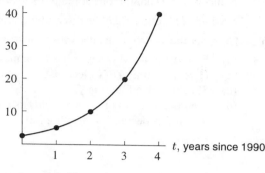

N, information packets per month (billions)

Figure 0.6

(a) Why do you think this graph is exponential?

(b) Suppose the graph is described by

$$y = 2.5e^{0.693t}.$$

When will the 100 billion mark be exceeded?

(c) Give an expression for the instantaneous rate at which the number of messages increases at $t = 2$ in terms of a limit.

(d) Using a calculator, find a value for the expression in part (c), accurate to 3 decimals.

(e) Find the average rate at which the number of messages increased per month during 1990.

58. The function f has only one turning point and f'' is shown in Figure 0.7. Sketch f and indicate a, b and c on your sketch.

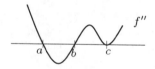

Figure 0.7

59. Differentiate the following with respect to the given variable:

(a) $y = \dfrac{1}{2\sqrt{t}} + e^2$, $\quad (t)$

(b) $y = \left(t^2 + t\right)^4 \ln(3t)$, $\quad (t)$

(c) $y = \dfrac{10x}{1 + 3e^{-x}}$, $\quad (x)$

(d) $y = 10^{0,203x} + (0,203x)^{10}$, $\quad (t)$

(e) $y = \text{bgsin}(5x) = \arcsin(5x)$, $\quad (x)$

(f) $x = \dfrac{4}{y} + 2$, $\quad (x)$

60. Say in each case for what well-known function, $y = f(x)$, the following holds:

(a) At every point, the rate at which the function changes is given by the y-coordinate.

(b) At every point, the rate at which the function changes is given by the reciprocal of the x-coodinate.

61. An animal population is described by

$$P(t) = 4000 + \frac{500}{t + 1} \sin\left(\frac{\pi}{6}(t - 3)\right)$$

with t measured in months, where $t = 0$ represents the beginning of the year. At what rate does the population decrease/increase at the beginning of the year?

62. A function f is continuous at the point $x = 1$ if $\lim\limits_{x \to 1} f(x) = f(1)$. Give three possible ways in which a function f can be discontinuous at $x = 1$. Three sketches will be accepted.

63. Consider the sketch of the derivative function f' in Figure 0.8.

    (a) Find the point(s) where f has local maximum or minimum value(s).
    (b) Give the point(s) of inflection of f.
    (c) Sketch such a function f.
    (d) Find the value(s) of x for which f is increasing most rapidly. Find the value(s) of x for which f is decreasing most rapidly.

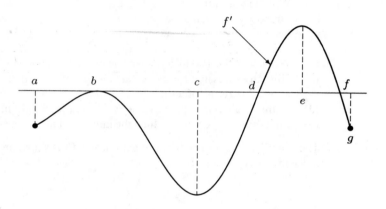

Figure 0.8

64. The number of cells, $N(t)$, in a culture, t minutes after a growth substance was added, is given by

$$N(t) = \frac{20t}{1 + t^2} + 100, \quad 0 \le t \le 5.$$

    where $N(t)$ is measured in thousands. Then, $N'(t)$ si given by

$$N'(t) = \frac{20(t^2 - 1)}{(t^2 + 1)^2}.$$

    Complete the following:
    (include units where applicable)
        Initially, the number of cells _____ at a rate of _____.
        At $t = $ _____, a critical point is reached, and by applying the first derivative test as follows:

It is clear that the function reaches a _____ in this point. Thereafter, the number of cells _____. The function has a global minimum where $t =$ _____, namely _____ and a global maximum, where $t =$ _____, namely _____. If

$$N''(t) = \frac{40t(3 - t^2)}{(t^2 + 1)^3},$$

it is clear that the function has point(s) of inflection in $t =$ _____.

65. Table 0.5 which shows the number of bags of oranges sold in one month, $f(p)$, against the price p per bag (in cents).

   (a) Find an approximation for $f'(900)$.
   (b) Describe what is meant by $f'(800) = -16$.

TABLE 0.5

Price p (in cents)	750	800	850	900	950
Number of bags, $f(p)$	50,000	48,000	44,000	37,000	29,000

66. Find $\dfrac{dy}{dx}$ if
   (a) $y = x^e \cdot (1.012)^x$
   (b) $y = \ln\left(\dfrac{x}{2} + \dfrac{2}{x}\right)$
   (c) $y = \sqrt{e^x + e^{-x}}$

67. A flu epidemic spreads amongst a group of people according to the formula

$$P(t) = \frac{1000}{1 + 199e^{-0.8t}} \qquad t \text{ is days}$$

where $P(t)$ represents the number of people that are infected on day t.

   (a) How many people are infected on the fifth day?
   (b) At what rate do the people become infected on day 5?

68. (a) Sketch the graph of a continuous function f for which the following holds:
       at $x < 0$, the function f increases at an increasing rate
       at $x > 0$, the function f increases at a decreasing rate.

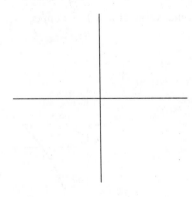

Figure 0.9

124

(b) Sketch the graph of f' if the graph in Figure 0.10 represents f:

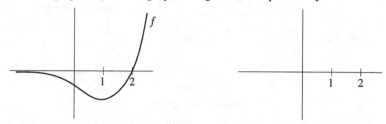

Figure 0.10   *Figure 0.11*

(c) Sketch the graph of a function f given the information in Figure 0.12. Mark x_1 and x_2 on your graph of f.

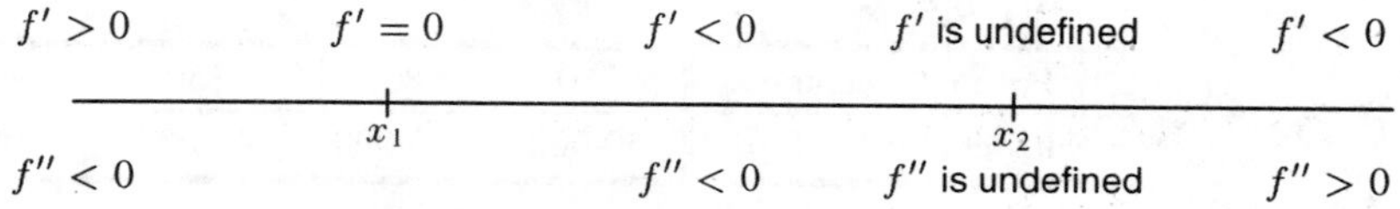

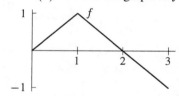

$f' > 0$ $f' = 0$ $f' < 0$ f' is undefined $f' < 0$

x_1 x_2

$f'' < 0$ $f'' < 0$ f'' is undefined $f'' > 0$

Figure 0.12

(d) Find $F(3)$ if $F' = f$ and $F(0) = 0$ and the graph of f is in Figure 0.13.

Figure 0.13

69. Consider the function $f(x) = \ln x$ as in the sketch. Suppose an approximation for $\int_1^3 \ln x\, dx$ is obtained by using Riemann-sums.

(a) Take $\Delta t = 0.5$ and shade the area that represents the left-hand sum on the sketch in Figure 0.14.

(b) Find a lower estimate for $\int_1^3 \ln x\, dx$. (Take $\Delta t = 0.5$).

(c) For what value of n will the upper and lower estimates differ by less than 0.1?

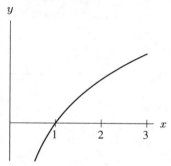

Figure 0.14

70. Evaluate

    (a) $\displaystyle\int_0^1 (\sqrt{x} + e^{2x+1})\,dx$

    (b) $\displaystyle\int x^2(x^3 + 4)^3\,dx$

    (c) The area included between the curve $y = e^x - 1$ and the x-axis between $x = -1$ and $x = 1$. (See Figure 0.15.)

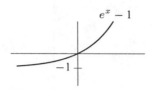

Figure 0.15

71. The size of an animal population varies according to

$$P(t) = 4000 + 500\cos\frac{\pi}{6}t$$

    where t is in months from the beginning of the year.

    (a) What is the average rate at which the animals increase/decrease over the first three months of the year?
    (b) What is the average number of animals in the herd during the first three months of the year? Use an integral.

72. Sketch f given that

$$f(x) = \frac{x}{(x+1)^2} \qquad f'(x) = \frac{1-x}{(x+1)^3} \qquad f''(x) = \frac{2x-4}{(x+1)^4}.$$

    Do this by finding:

    (a) The intercepts and the asymptotes
    (b) The critical points, the results of first-derivative test, and where f is increasing or decreasing
    (c) The results of the second-derivative test, and the points of inflection.
    (d) Sketch the graph.

73. The function $f(x) = \dfrac{x^3}{3} - \dfrac{x^2}{2} - 6x$ has critical points at $x = -2$ and $x = 3$. Obtain an estimate for the first (smallest) point of intersection of f with the x-axis by making a rough sketch of f. Apply Newton's method twice to improve this estimate.

74. A brick is heated in an oven and taken out to cool off after a certain time. The temperature T of the brick at any time t is given by

$$T = 10e^{-(t-1)^2}, \quad t \geq 0.$$

    T is measured in °C and t in minutes.

    (a) What is the temperature of the brick when it is placed in the oven?
    (b) At what time t is the brick taken out of the oven?
    (c) What will the temperature of the brick eventually be?
    (d) Sketch the temperature/time curve.

75. A mirror is formed by joining a half-circle to a rectangle, as in Figure 0.16. If the total circumference of the mirror is 200 cm, what should the radius of the half circle be for the area of the mirror to be a maximum?

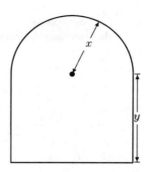

Figure 0.16

76. Determine $\int t^2 \sin t \, dt$.

77. Given

$$\int \frac{cx + d}{(x - a)(x - b)} \, dx = \frac{1}{a - b} \left[(ac + d) \ln |x - a| - (bc + d) \ln |x - b| \right] + C$$

determine

$$\int_{-1}^{1} \frac{3x + 1}{x^2 - x - 6} \, dx$$

78. Given the curve

$$y = ax^3 + bx^2 + cx + d, \quad a \neq 0,$$

find the relation between the parameters a, b, and c that will ensure that the curve

(a) has only one turning point.
(b) has no turning points.

79. The sketch in Figure 0.17 represents the function f with $f(x) = e^{-ax} \sin x$, $a > 0$ and $x \geq 0$.

(a) Determine the x-intercepts of f.
(b) What must the value of a be so that $x_0 = \dfrac{\pi}{4}$?
(c) If $x_0 = \dfrac{\pi}{4}$, calculate x_1.

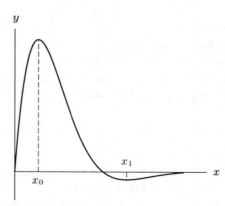

Figure 0.17

80. The demand for cold drink on campus is D liters per day. This demand is a function of the price of cold drink c (in dollars/liter) and the price of milk m (dollars/liter) and is given by
$$D(c, m) = 200c - \frac{500}{m}.$$

    (a) Use the same axis of reference to represent graphically the demand for cold drink as a function of the price of cold drink per liter if the price for milk is \$2, \$2.50, and \$3 respectively.

    (b) How do the prices of cold drink and milk influence the demand for cold drink?

81. Supose $Q = te^{-bt}$ with $t \geq 0$ and b a constant.

    (a) For which values of t will Q increase and for which values of t will Q decrease?

    (b) Determine the coordinates of the point of inflection of Q.

    (c) What happens to Q as t increases indefinitely?

    (d) Sketch the graph of $Q = te^{-bt}$ and show how the value of the constant b influences the shape of the graph.

    (e) If Q represents the level of alcohol in a person's blood, what could the symbols b and t represent?

82. With x people on board an airline makes a profit of $(900 - 3x)$ rands per person for a specific flight.

    (a) How many people would the airline prefer to have on board?

    (b) What is the maximum number of passengers that can board such that the airline still profits?

83. Apply Newton's method to improve one on 5 as an approximation for the square root of 27.

84. (a) Represent the functions f and g on the same axes in 3-space, where
$$f(x, y) = 2 - \sqrt{x^2 + y^2}, \quad 1 \leq f(x, y) \leq 2$$
    and
$$g(x, y) = x^2 + y^2, \quad 0 \leq g(x, y) \leq 1.$$

    (b) Sketch the domain of f in the xy-plane.

85. $f(x)$ is the age of antarctic ice (in hundreds of years) at a depth of x meters below the surface.

    (a) Give in words the practical meaning of the equation $f(10) = 15$.

    (b) Give in words the practical meaning of the equation $f^{-1}(20) = 12$.

    (c) Is f increasing or decreasing and why?

86. Solve for t:
$$75 \cdot 14^t = 50 \cdot 12^t$$

87. Solve for the functions below:

    (a) $g(t) = 15 \cdot 10^t$ Answer using the form $A \cdot e^{kt}$.

    (b) $h(t) = 100 \cdot e^{-0.3t}$ Answer using the form $P_0 \cdot a^t$.

88. A population of rabbits is growing. In 1996 there are 10,000,000 rabbits, and the increase is 20% per decade (per ten years). What will the population be in the year

    (a) 2006?

    (b) 2096?

    (c) 1997?

    (d) Write a formula for the population t years from now.

89. (a) Give a formula for the exponential saturation function $k(t)$.

    (b) Determine t such that $k(t) = 99$.

128

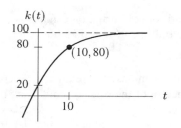

Figure 0.18

90. (a) Sketch a graph of $f'(x)$ on the axis provided if the graph of $f(x)$ is in Figure 0.19.

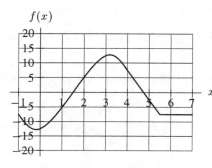

Figure 0.19   *Figure 0.20*

(b) Using the graphs in part (a), complete the tables of values:

TABLE 0.6

x	0	10	20	30	40	50	60	70
$f(x)$								

TABLE 0.7

x	10	20	30	40
$f'(x)$				

(c) Determine if the following are positive, negative or zero, and explain how you know.

(a) $f''(5)$ (b) $f''(25)$ (c) $f''(45)$

91. If $g(v)$ is the fuel efficiency of a car moving at v miles per hor (efficiency measured in miles per gallon):

(a) Give the meaning, in plain English, of the equation $g(55) = 27$.
(b) Give the meaning, in plain English, of the equation $g'(55) = -0.54$.
(c) Give the units for $g'(v)$.
(d) Why is $g'(55)$ negative?

92.

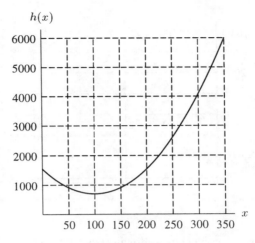

$h(x)$

50 100 150 200 250 300 350

Figure 0.21

(a) For the function in Figure 0.21, give a $N = 5$ term left Riemann sum approximation:

$$\int_{100}^{350} h(x)\,dx \approx \underline{\qquad} + \underline{\qquad} + \underline{\qquad} + \underline{\qquad} + \underline{\qquad}$$

$$= \underline{\qquad}$$

(b) Shade the area you actually computed in part (a).
(c) Is your approximate answer in part (a) too high or too low? Why?
(d) Give the $N = 5$ term right Riemann sum for the inverval in (a):

$$\int_{100}^{350} h(x)\,dx \approx \underline{\qquad} + \underline{\qquad} + \underline{\qquad} + \underline{\qquad} + \underline{\qquad}$$

$$= \underline{\qquad}$$

(e) The average of $h(x)$ over the range $100 \le x \le 350$ is approximately \underline{\qquad}.
(f) Draw in the horizontal line on the graph that corresponds to the average of $h(x)$ over the range $100 \le x \le 350$.

93. (a) Give a 4-term left Riemann sum approximation for:

$$\int_{30}^{3} 3\sqrt{x+1}\,dx \approx \underline{\qquad} + \underline{\qquad} + \underline{\qquad} + \underline{\qquad} = \underline{\qquad}.$$

(b) Is your answer in part (a) an overestimate or an underestimate? \underline{\qquad}
(c) If you used a 500-term Riemman sum in part (a), how big would the difference between the right and the left sums be? $|RRS - LRS| = \underline{\qquad}$.

94. (a) Sketch the graph of $g(x)$ given the graph of $g'(x)$. Assume $g(0) = 2000$.

$g(x)$

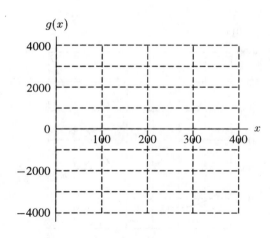

$g'(x)$

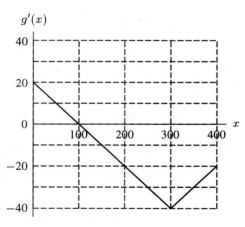

(b)   Complete the following tables of values:

TABLE 0.8

x	0	100	200	300	400
$g(x)$	2000				

TABLE 0.9

x	100	200	300	400
$g'(x)$				

(c)   Determine whether each of the following is positive or negative:

(a) $g(50)$ _____   (b) $g(150)$ _____   (c) $g(350)$ _____
(d) $g'(50)$ _____   (e) $g'(150)$ _____   (f) $g'(350)$ _____
(g) $g''(50)$ _____   (h) $g''(150)$ _____   (i) $g''(350)$ _____

(d)   What is happening on the graph of $g(x)$ at

(i) $x = 100$?

(ii) $x = 300$?

95.   Find the derivavtives.

(a) $y = (x^2 + 3\sqrt{x})^5$
(b) $y = 8 \sin(2^{-x})$
(c) $y = 12e^{\tan(x)}$
(d) $y = x \cos(3x)$
(e) $y = \dfrac{\ln x + 5}{x^2 + 7}$

96.   Find the equation of the line tangent to the curve $y = \sqrt{3x + 7}$ at the point above $x = 3$.

97.   Compute $\dfrac{dy}{dx}$

(a) $y = 10^x + 18 \arctan(x^2)$
(b) $y = x \cos(\ln x)$
(c) $y = \dfrac{x^2 + e^{3x}}{\sin(2x) + 8}$

98.   Find the equation of the lines tangent to the curve $2^x + yx + y^3 = 12$ at the point where $x = 1$ and $y = 2$.

99.   (a)   Sketch the graph of $y = 2x^3 + 3x^2 - 36x + 100$.

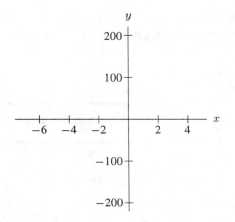

(b) The graph is:
    (i)   Increasing and concave up on the interval _____.
    (ii) Increasing and concave down on the interval _____.
    (iii) Decreasing and concave up on the interval _____.
    (iv) Decreasing and concave down on the interval _____.

100. Find possible formulas for $g(t)$ and $k(t)$.

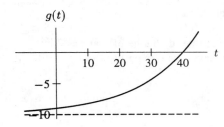 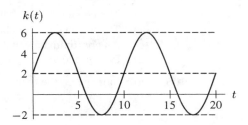

$g(t) =$ _____ $k(t) =$ _____

101. (a) Write the following function in the form Ae^{kt}

$$a(t) = 40(0.8)^t = $$ _____

    (b) Write the following function in the form $P_0 a^t$

$$b(t) = -10e^{1.02t} = $$ _____

102. Sketch the graph of a function which is increasing and concave down.

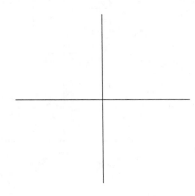

103. Suppose values for $g(x)$ are given in Table 0.10. Do you think that $g(x)$ concave up or concave down? How do you know?

TABLE 0.10

x	1	2	3	4	5	6
$g(x)$	100	90	81	73	66	60

104. Suppose that $S(q)$ is the price per unit (in $) of widgets which will induce producers to supply q thousand widgets to the market, and suppose that $D(q)$ is the price per unit at which consumers will by q thousand items.

    (a) Which is larger, $S(100)$, or $S(150)$, and why?
    (b) Which is larger, $D(100)$, or $D(150)$, and why?
    (c) If $D(100) = 10$ and $S(150) = 10$, what will you predict about the future selling price of widgets (currently at $10)?

105. Suppose $f(t) = t^2 + t$.

    (a) The change in $f(t)$ between $t = 2$ and $t = 5$ is _____.
    (b) The average rate of change in $f(t)$ between $t = 2$ and $t = 5$ is _____.
    (c) An approximate value for the rate of change of f at $t = 2$ is _____.
    (d) How could you improve your estimate in part (c)?

106. (a) Indicate on the graph:

        (i) A line segment whose length equals the change Δh in $h(x)$ between $x = 20$ and $x = 40$.

        (ii) A line segment whose slope equals the average rate of change $\dfrac{\Delta h}{\Delta x}$ of $h(x)$ between $x = 20$ and $x = 40$.

        (iii) A line whose slope equals the deirvative $h'(10)$.

        (iv) A point on the graph where $h' = 0$.

    (b) $h'(30) \approx$ _____.

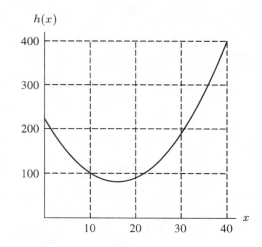

107. You are housing minister (in the year 1996) for a country with 30 million people. You have been asked to predict the population 5 years and 10 years from 1996, as part of a 10 year master plan for housing. Census records show that the population was 22.684 million in 1986 and 26.087 million in 1991. Give your predictions, with justification in words.

108. (a) Let $f(x) = x^2 + 3$. Derive an exact formula for the derivative $f'(x)$ by computing algebraically the limit of a difference quotient.
      (b) Write an equation for the line tangent to the graph of $y = x^2 + 3$ at the point where $x = 1$.

109. Compute the derivatives of the following functions:

      (a) $y = 2x^4 + \dfrac{x^2 - 10x}{x^3}$

      (b) $y = 3.5^x + \dfrac{2}{\sqrt{x}}$

      (c) $y = 8e^{-3x}$

110. (a) Sketch the graph of $g(x)$, given the graph of $g'(x)$. Assume $g(0) = -1000$.

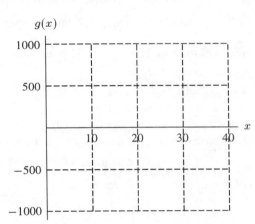
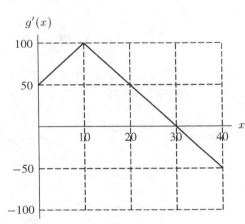

      (b) Complete the following tables of values:

TABLE 0.11

x	0	10	20	30	40
$g(x)$	-1000				

TABLE 0.12

x	10	20	30	40
$g'(x)$				

      (c) Determine whether each of the following is positive or negative.
           (a) $g(5)$_____   (b) $g(25)$_____   (c) $g(35)$_____
           (d) $g'(5)$_____   (e) $g'(25)$_____   (f) $g'(35)$_____
           (g) $g''(5)$_____ (h) $g''(25)$_____ (i) $g''(35)$_____
      (d) What is happening on the graph of $g(x)$,
           (i) at $x = 10$?
           (ii) at $x = 30$?

111. You plan to approximate the definite integral $\displaystyle\int_5^{10} (2x + 3)^2 \, dx$ by Riemann Sums.

      (a) Which will be larger, the Right Riemann Sum or the Left Riemann Sum, and how do you know?

(b) Compute the difference |Right Riemann Sum − Left Riemann Sum| if you use N terms, with:

   (i) $N = 100$. |RRS − LRS| = _____

   (ii) $N = 1000$. |RRS − LRS| = _____

   (iii) $N = 10{,}000$. |RRS − LRS| = _____

(c) Would N terms be sufficient to compute $\int_5^{10} (2x+3)^2 \, dx$ to the nearest integer, if

   (i) $N = 100$?

   (ii) $N = 1000$?

   (iii) $N = 10{,}000$?

112. There is a population of $P(t)$ thousnad bacteria in a culture at time t hours after the beginning of an experiment. You know tha t$P(10) = 20$, $P'(10) = 0.4$, and $P''(10) = 0.008$.

(a) Give the meaning, in practical terms, of the equation $P'(10) = 0.4$.

(b) Using the value of $P'(10)$, make a prediction for the population at time $t = 10.5$ hours.

(c) Give the meaning, in practical terms, of the equation $P''(10) = 0.008$. Would you expect the population to increase more or less between $t = 10.5$ and $t = 11$ than between $t = 10$ and $t = 10.5$?

(d) Using the value of $P''(10)$, make a prediction for the rate of change of population at time $t = 10.5$.

(e) Give your best prediction of the population at $t = 11$ hours.

113. Consider the following statements:

(a) If $f'' < 0$, then f' is decreasing.

(b) $y = x^{-1} + 3$ is a polynomial.

(c) If $MR < MC$, then less items should be manufactured.

(d) The ln-function will dominate the polynomial $y = x^{0.1}$ in the long run.

(e) The function $P = \dfrac{100}{1 + 20e^{-0.03t}}$ grows at a rate of 3% for small values of t.

Give only the number(s) of the statement(s) that is(are) *true*.

114. Find $\dfrac{dy}{dx}$ if:

(a) $y = e^{-0.01x} + \ln x$

(b) $y = 4x^2 \sin(\pi x)$

(c) $y = (2^x + \sqrt{x})^5$

115. Sketch, if possible, a function f with the given properties. Explain if this is impossible.

(a) $f(x) > 0$, $f'(x) > 0$ and $f''(x) < 0$ for all real numbers x.

(b) $f(x) > 0$, $f'(x) > 0$ and $f''(x) > 0$ for all real numbers x.

116. The cost to produce q aircraft is given by the function

$$C(q) = 2.5q^{0.848}$$

with C in millions of rands. Find the marginal cost for $q = 50$ and interpret your answer.

117. If 100 is invested at r% interest per year, compounded yearly, then the yield after 15 years is given by

$$F = 100(1 + \frac{r}{100})^{15}.$$

(a) Find $\dfrac{dF}{dr}\Big|_{r=5}$.

(b) Interpret your answer in part (a).

118. Bank A offers 12% interest, compounded yearly and bank B offers 11.8% interest, compounded continuously.

    (a) You want to invest $1000 for 10 years. Which bank should you choose? Show your computations.

    (b) What is the effective interest rate of bank B?

119. A person breathes every three seconds. The volume of air in his lungs varies from 2 to 4 litres.

    (a) Which of the following functions describes the situation best?

        (i) $y = 2 + 2\sin(2\pi t/3)$

        (ii) $y = 3 + \sin(\pi t/3)$

        (iii) $y = 3 + \sin(2\pi t/3)$

        (iv) $y = 2 + 2\sin(6\pi t)$, with t in seconds.

    (b) At what rate is the volume of air in his lungs changing after 1 second?

120. A biologist found that the number of Drosophila fruit flies, $N(t)$, assumes the following growth pattern if the food source is limited:

$$N(t) = \frac{400}{1 + 39e^{-0.4t}}.$$

    (a) How many fruit flies were there in the beginning?

    (b) At what time was the population increasing most rapidly?

    (c) At what rate does the number of fruit flies increase after three days?

121. Consider the two concentration curves, A and B, shown in Figure 0.22.

    (a) For which value of C (approximately) will A maintain a minimum concentration of C ng/ml for exactly two hours?

    (b) Determine the equation of curve B.

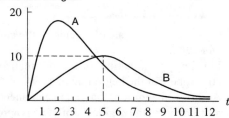

concentration in ng/ml

Figure 0.22

122. Sketch the curve $y = 10(1 - e^{-t})$.

123. (a) Complete:

        (i) $\dfrac{d}{dx}(e^x \cdot \ln x) =$

        (ii) $\dfrac{d}{dx}\left(\dfrac{10}{1 + 9e^{-x}}\right) =$

    (b) What does the expression $\dfrac{f(2) - f(1.99)}{0.01}$ represent if f is a function?

    Complete:

    The instantaneous rate of change of a function f in a point x is defined as:

    (c) Sketch a function f with the following properties:

        (i) $f'(x) > 0$ and f' increasing for $x > 0$.

    (ii) $f'(x) > 0$ and f' decreasing for $x < 0$.

  (d) Sketch the graph of f' if the graph in Figure 0.23 represents the function f:

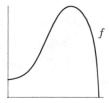

          Figure 0.23   *Figure 0.24*

  (e) Complete:

        If C is a cost function and $\int_{40}^{50} C'(q)\,dq = 55$, it means that _____

  (f) Complete:

        If $f'(t) = 2t$ is a production rate, measured in items/hour, then _____

        which means that _____.

124.   The amount of carbon dioxide in the atmosphere, measured in parts per million at the beginning of every fifth year since 1970, is given in the following table:

Year	CO_2 (in dpm/in ppm)
1970	325.3
1975	331.0
1980	338.5
1985	345.7
1990	354.9

  (a) Give the average rate of change of the amount of CO_2 in the atmosphere between 1970 and 1980.

  (b) Find an approximation for the instantaneous rate of change of the amount of CO_2 in the atmosphere at the beginning of 1980.

  (c) Assume, in addition, that the instantaneous rate of change of the amount of CO_2 in the atmosphere at the beginning of 1985 is approximately 1.44 ppm/year and at the beginning of 1990 approximately 1.66 ppm/year. What can you conclude from this information and your answer in part (b) regarding the concavity of the curve in the eighties?

125.   Consider the cost function $C(q) = (q - 12)^3 + 120$ and the revenue function $R(q) = 10q$, with q the quantity in thousands and R the cost in thousands of rands.

  (a) Sketch the cost function.

  (b) Why is $q = 12$ a significant quantity?

  (c) Explain why the profit is at a maximum when $q = 13.826$.

126.   The size of an impala population is represented by the function $R(t) = 10 + 2\cos(\pi t/6)$, with t in months since the beginning of the year and where $R(t)$ is measured in thousands.

  (a) At what rate does the population change after 3 months?

  (b) Give the expression for calculating the average number of impala during the first three months. (You need not do any calculations.)

127.   Find the area included between the curves $y = \sqrt{x}$ and $y = \dfrac{1}{x}$ between $x = 1$ and $x = 2$.

128. Of a certain function G we know that: $G(1) = 0.29793972$, $G(0.1) = 0.03936165$, $G(0.01) = 0.00394929$, $G(0) = 0$.

    (a) Use these values to determine $G'(0)$, to as many decimals as you can confidently give an opinion on.

    (b) In addition, say that $G(0.001) = 0.000394942$. Can you improve on your answer in part (a)? Explain.

129. The two graphs in Figure 0.25 represents the inflow-rate and outflow-rate of a reservoir over a period of 8 months.

    (a) Was the volume of water in the reservoir increasing or decreasing during the first month?

    (b) When was the volume of the water in the reservoir at a maximum?

    (c) Was the volume of the water in the reservoir at the end of the eight-month period more or less than at the beginning? Explain.

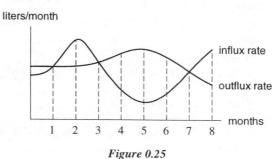

Figure 0.25

130. Consider the graph of the function $f(x) = e^{-x^2}$ in Figure 0.26.

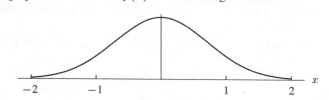

Figure 0.26

    (a) Approximate $\int_0^{1.5} e^{-x^2} dx$ by using a left-hand sum with 3 subdivisions.

    (b) Is your answer to part (a) a lower estimate or an upper estimate?

131. In the F & T Weekly of April 4, 1977, various aspects of the big cities in South Africa are compared. The following statement is made:

    House prices in Cape Town have risen at nearly double the pace of those in most other cities.

   Assume two houses (one in Cape Town and one in Pretoria) were both worth $200,000 five years ago. Study the symbolic explanation contained in lines (i), (ii), (iii).

     (i) $f'(x) = 2g'(x)$

     (ii)   So $\int_0^5 f'(x)dx = 2\int_0^5 g'(x)dx$

     (iii)   So $f(5) - f(0) = 2[g(5) - g(0)]$

    (a) What do the functions f and g respectively represent?

    (b) To which of the symbolic statements does the quote above refer?

    (c) Give the statement in line (ii) in words.

    (d) Use number values to illustrate what is meant by line (iii).

132.

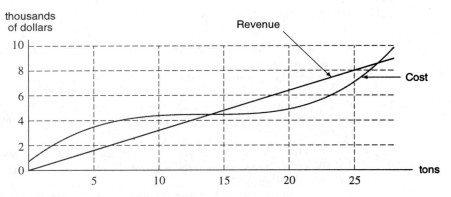

Figure 0.27

Above are cost and revenue functions for a certain chemical important in **industry.**

(a) It costs _____ to produce 10 tons.
(b) Revenue from sale of 10 tons is _____.
(c) Break even points are at _____ and _____.
(d) Marginal cost at 20 tons is _____ dollars/ton.
(e) Marginal revenue at 20 tons is _____ dollars/ton.
(f) Sale price is _____ dollars/ton.
(g) Should the company increase production beyond 20 tons? _____. **Explain** your reasoning.
(h) To maximize profit, the company should produce and sell _____ **tons, and** then its profit will be _____ dollars.

133. Suppose the graph of f is in Figure 0.28. Are the following **quantities positive, negative, or** zero?

(a) $f(A)$ (b) $f'(A)$ (c) $f''(A)$ (d) $f(B)$
(e) $f'(B)$ (f) $f''(B)$ (g) $f(C)$ (h) $f'(C)$
(i) $f''(C)$ (j) $f(D)$ (k) $f'(D)$ (l) $f''(D)$

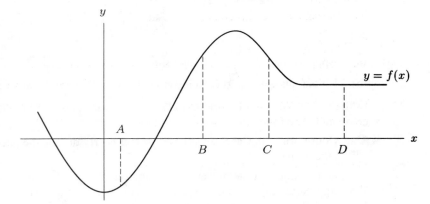

Figure 0.28

134. Suppose $g(t)$ is the height in inches of a person who is t years old.

(a) Give a reasonable approximation for

(a) g(0)   (b) g(30)

(b) Give the meaning, in plain English, of $g'(10) = 2$.
(c) What is $g'(40)$?

135. Find the derivatives:
   (a) $y = 4x^3 + 10x + 1$
   (b) $y = 12 \cdot 10^x$
   (c) $y = 3\sqrt{x}$

136. Show how to use difference quotients to derive the formula for the derivative of $f(x) = x^2$.

137. Write the equation of the line tangent to the curve $y = x^2 + 2x$ at the point where $x = 3$.

138. The population of Mexico t years after 1980 is given by the formula $67.4(1.02)^t$ million. At what rate was the population growing in 1996?

139. Find possible formulas for the functions f and g:

   TABLE 0.13

x	1	2	3	4	5	6
$f(x)$	0.500	0.750	0.875	0.938	0.969	0.984

   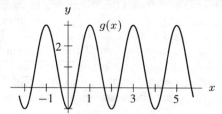

   Figure 0.29

140. A new music company wantes to start selling compact discs. The profit π (in thousands of dollars) is $\pi(p) = 160p - 6.4po^2$ is p is the price of a compact disc (in dollars).
   (a) Sketch the graph of $\pi(p)$, showing the zeros of the function.
   (b) Find the coordinates of the maximum point on the graph.
   (c) Interpret the maximum point and its coordinates in practical economic terms.

141. A new species, introduced into an environment in which it has no natural predators, grows exponentially with continuous growth rate $k = 0.095$ per year.
   (a) If there were initially 45 individuals introduced, write the formula for the number of individuals after t years.
   (b) What is the approximate doubling time of this population?
   (c) About how long after introduction will the population reach 250 individuals?

142. The world's only manufacturer of left-handed widgets has determined that if q left-handed widgets are manufactured and sold per year, then the cost function is

   $$C = 8000 + 40q$$

   and the manufacturer's revenue function is

   $$R = pq$$

   if the price is p. The manufacturer also knows that the demand function for left-handed widgets is

   $$q = 2000 - 25p$$

   at price p.
   (a) Using the demand function, rewrite the cost and the revenue functions in terms of price p.
   (b) Write the profit function π in terms of price p and sketch its graph.
   (c) For about what price is the profit largest? How many left-handed widgets should be produced at that price?

143. Your friend Herman operates a neighborhood lemonade stand. He asks you to be his financial advisor and wants to know how much lemonade he can make with the $3.27 he happens to have on hand. The only information he can give you is that once last month he spent $2 and made 19 glasses of lemonade, and another time he spent $5 and got 83 glasses of lemonade. You decide to use this data to create a cost function, $C(q)$, giving the cost in dollars of making q glasses of lemonade.

    (a) You first decide to create a linear cost function based on this data. What is the linear cost function, and how much lemonade can Herman make according to this model?
    (b) You decide to create an exponential cost function. What is the exponential cost function, and how much lemonade can Herman make according to this model?
    (c) Herman routinely sells his lemonade for 10¢ per glass. In the case of the linear model, what is his break-even point?
    (d) In the case of the exponential model, what can you tell Herman about maximizing his profit? Again assume a sales price of 10¢ per glass.

144. Your rich eccentric friend has hired you to cover his back yard with grass and patio stone. Instead of drawing a map, he has given you equations of the boundary lines. If the southwest corner of his yard is taken as the origin, with the x-axis pointing eastward and distances measured in feet, then the boundaries of the yard are the lines $x = 0$, $x = 100$, $y = 0$, and $y = 110 - 0.5x$. The border between grass and stone is $y = 40 + 20\cos(\pi x/40)$, with grass covering all of the yard south of the curve. This border also bounds one side of his pool, the other side of the pool being surrounded by the curve $y = 60 - 0.05(x - 40)^2$. All the rest of the yard is to be covered with stone.

    (a) Sketch a map of your friend's yard.
    (b) Estimate, to the nearest square yard, the area of (i) the grass and (ii) the stone.

145. The graph in Figure 0.30 is the graph of $N = C(t)$, the cumulative number of customers served in a certain store during business hours one day, as a functions of the hour of the day.

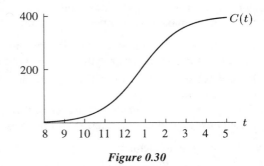

Figure 0.30

    (a) About when was the store the busiest?
    (b) What does $C'(t)$ mean in practical terms?
    (c) Estimate $C'(11)$.
    (d) Let $t = C^{-1}(N)$ be the inverse function of C. What does $C^{-1}(250)$ mean? Estimate $C^{-1}(250)$.
    (e) What are the units of $d/dN\left(C^{-1}(N)\right)$. Estimate $d/dN\left(C^{-1}(N)\right)$ at $N = 250$?